The Beginner's Cut Flower Garden

Amagansett Free Library
215 Main St, PO Box 2550
Amagansett, NY 11930
www.amagansettlibrary.org

The Beginner's Cut Flower Garden

Grow, Nourish, and
Create Bliss Year-Round

Elizabeth Brown

Photography by Lindsay Fairchild

Timber Press
Portland, OR

Copyright © 2026 by Elizabeth Brown. All rights reserved.

Hachette Book Group supports the right to free expression and the value of copyright. The purpose of copyright is to encourage writers and artists to produce the creative works that enrich our culture. The scanning, uploading, and distribution of this book without permission is a theft of the author's intellectual property. If you would like permission to use material from the book (other than for review purposes), please contact permissions@hbgusa.com. Thank you for your support of the author's rights.

Timber Press
Workman Publishing
Hachette Book Group, Inc.
1290 Avenue of the Americas
New York, New York 10104
timberpress.com

Timber Press is an imprint of Workman Publishing, a division of Hachette Book Group, Inc. The Timber Press name and logo are registered trademarks of Hachette Book Group, Inc.

Printed in Dongguan, China (TLF), on responsibly sourced paper

Text design, cover design, and floral illustration by Sara Isasi

The publisher is not responsible for websites (or their content) that are not owned by the publisher.

The Hachette Speakers Bureau provides a wide range of authors for speaking events. To find out more, go to hachettespeakersbureau.com or email hachettespeakers@hbgusa.com.

ISBN 978-1-64326-501-8

A catalog record for this book is available from the Library of Congress.

To my husband and children, every flower is for you.

Contents

11 Introduction
16 How to Use This Book

Winter
Arrive & Inspire
—— 19 ——

20 A Flower Is a Memory
21 Finding Inspiration
30 Flowers for Cutting
32 A Mix of Flower Varieties
54 Sourcing Flowers
55 Starting Your Flowers from Seed
66 Designing Your Cutting Garden
68 How to Create a Garden Rhythm

Spring
Awake & Prepare
—— 73 ——

78 Basic Garden Supplies
84 Developing a Watering Plan
89 Preparing Your Garden
96 The Beginning of Flowering
97 Identifying Your Favorite Spring Bulbs
105 Adding Shrubs
108 Hardening Off Your Seedlings
112 Harvesting Spring Flowers
121 The Importance of Remaining Present

Summer
Harvest & Design
——— 125 ———

128 A Beginner's Guide to Floral Design
142 Tending the Summer Garden
155 Taking Time to Connect
161 Considering Natural Dyes
167 Slowing Down in the Garden

Autumn
Closing & Reflection
——— 173 ———

175 Autumnal Rhythms
183 A Word on Season Extension
185 It's Time for the Dahlia Dig
190 Cleaning Your Tools
191 Creating a New Garden Bed for Spring
195 Working with Dried Flowers
208 Embracing and Creating Tradition
220 A Time for Reflection

222 References
224 Acknowledgments
226 Index

Introduction

A garden can be whatever you want it to be. It does not have to be rows of flowers. It does not have to be well-timed and neatly manicured. A garden can be a single bloom in a pot, a houseplant, or an herb on a kitchen windowsill—just a bit of green that pulls you from your busy mind, and back to earth. A single cut flower, like the flicker of a candle, can expand a room with unexpected beauty.

I fell into our garden, or was rather pushed into gardening, by forces unknown. There was no foresight, no ordering of tools, no grand declaration. Just the simple repetitive act of dropping my knees to the ground and becoming present with the world in front of me. What started as a means of distraction during the pandemic turned into a full-blown obsession. My experience is not unique—many people found their way to a long-neglected rosebush or bit of earth because they were home, but also as a form of stress relief and self-resilience. This is not a novel concept. One only has to look as far back as the victory gardens of World War II to realize we return to the soil in times of difficulty.

Before the chaos and isolation of 2020, I had been slowly learning how to tend to a garden for years. But it was all a bit frenzied and soulless. I was overwhelmed by my lack of knowledge, my inability to differentiate between weed and flower, and the nagging line of thought that there were so many other things I should be doing. My body was in the garden, but my mind was always elsewhere. With two small children, a demanding full-time job, and a never-ending to-do list, there were many times I considered putting the garden to bed, permanently. The thought of this now makes me shudder. Each time I wandered the garden, full of weeds, my heart sank a bit, and I felt a failure. How would I ever find time to make the garden beautiful? How could I have killed yet another perennial flower?

But with nowhere to go in our state of social isolation, and with all the jigsaw puzzles solved and bread baked, I was able to return to the garden with a new mindset. Each morning, as the sun expanded upon my windowpane, the general sense of fear and loneliness was replaced with excitement about the possible transformation of the flower garden while I slept. Had the peonies opened? Would I learn the aphids had disappeared into the night, leaving my lupines to rest? A feeling of wonder emerged, and that feeling superseded those of unworthiness and fear.

Instead of floundering inside, attempting to teach my son to read and my daughter to do multiplication, we took to the dirt. Our days were filled with wildflower hunts, planting seedlings, and the smell of rain on the hyacinths. Together we learned how to identify sedum and wild geranium as they popped through the earth. Together we dug new garden beds and collected apple blossoms after the rain. And together, when it all was called off, the return to work and school upon us, it became quite clear we were different. No longer was I too busy for the garden; in fact, suddenly it seemed the only place I wanted to be. I was thinking like a child, and that made it easier to tap into my beginner's mind, without judgment. I could teach my children while also learning myself. There was a holistic element to this, a grounding element, that made it easier for me to engage in gardening as a newcomer. It's not easy, especially in a society that values exceptional talent and expertise. Is there room for *decent* gardeners? *Amateur* sort of stings as a word. And there's a bit of a stigma around being a beginner, but isn't the beauty about constantly learning? Isn't that what keeps us engaged, and in a state of wonder?

It was during this summer that I began thinking about a beautiful little book my grandmother kept in a cupboard in her living room. This book, *An Island Garden* by Celia Thaxter, had inspired my Grammy to plant lots of vibrant red poppies, among other flowers. She explained that the rocky soil on her little plot of land was presumably exactly the same as the soil of that in Celia's garden, which was on a windswept island off the coast of Maine. From our spot on the front porch, Grammy would slowly motion toward the ocean, where the tracings of Appledore Island hung low on the horizon. It was as if I could draw an invisible line, over the sea, six miles between Grammy's garden on the mainland and Celia's on the island. Some sort of bond had been born in the soil, between two women who had never met, but shared the same love of art and flowers, one hundred years apart. I felt this connection, and it spoke deeply to me.

This very same book arrived in the mail nearly twenty-five years later on my front porch, long after my grandmother taught me about pulling seaweed from the ocean and using it to fertilize the garden as Celia had. I devoured the book in a day, underlining moving passages and taking notes on flower types and organic pest-control methods (by this, I mostly mean hand pulling slugs at night). This book is entirely different from every other gardening book I have in my collection. It speaks in lyrical bits of poetry, describing what I often cannot—the way in which your soul attaches to the earth when you spend time there, how the life cycle of a flower can bring one to tears, or how the worries of the world seem to fade when you put your mind toward a poppy expanding, almost impossibly, in the morning sun. It is less about how to grow and more about how to *live*. How to slowly move about a tiny garden and marvel at each individual bloom. It was not lost on me that the amount of beauty Celia created through flowers, inspiring poetry and paintings, occurred in a small fifty by fifteen–foot plot. Every stem meticulously thought over and loved deeply. A stark contrast to more well-known inspirational gardens with acres of land, like that of Monet. Inspiring, for sure, but relatable? Less so.

While Celia's words spoke to the soul of the garden, my clinical brain needed to understand more of the science behind this state of euphoria. I felt it to be true, but *why* does one feel better in a garden? Was it perhaps the additional vitamin D, the physical movement, the microbes in the soil, a combination of all these things? This wasn't just all poetry and prose, and I knew I was not imagining how much better I felt between rows of flowers. There had to be something scientific to it.

Through this line of questioning, I stumbled upon the practice of therapeutic horticulture, and quickly discovered that the healing benefits of tending to or spending time in a garden have never been disputed. This is why so many large hospitals offer healing gardens to their patients. It's why my physician grandfather encouraged my grandmother to get her hands into the garden every day while she recovered from an infection following the birth of her twelfth child.

As we all pull ourselves out from trauma and loss, this fact is simple: we all might benefit from the simple act of sowing a seed in our backyard. While I would never suggest that gardening replace the treatment of a mental-health professional, taking time to move slowly in a garden can be an incredible life practice, a freeing one. After all, gardening has been proven to increase dopamine levels and lower stress.

Around the time I began my certification in therapeutic horticulture, I had a conversation with a member from a local garden club and pitched the idea of

speaking to their group on the mental-health benefits of tending to a garden. This idea was immediately rejected because, she explained, all gardeners already know this. And while this sentiment could be true, it made me wonder: why does it still feel like a well-kept secret, especially at a time when many people long to feel better and have not a clue how healing a garden might be? Any barriers need to be removed, at once, so the simplicity and doable-ness of gardening is obvious to all.

    This book is an attempt to reveal the secret many gardeners hold: that often, growing a garden has little to do with the flowers. Gardening makes you feel better just by putting your hands in dirt. Its bliss doesn't rest in the fleeting moment when one snips a bloom cultivated by our own hand. The beauty of the garden flows into your dreams and psyche. Your eyes, now accustomed to noticing the smallest bit of earth awakening in early spring, can spot the impossibility of a buttercup emerging from a crumbling bit of pavement. As Celia said so beautifully,

> *He who is born with a silver spoon in his mouth is generally considered a fortunate person, but his good fortune is small compared to that of the happy mortal who enters this world with a passion for flowers in his soul.*

You will likely notice I weave bits of practical information about creating a cut flower garden alongside reminders of how to find mindful moments in your garden. Like many gardeners before me, I find it difficult to discuss the process of creating a garden without tapping into the more spiritual elements of the experience. But it's this that I often find missing in gardening books, and why I continue to turn to Celia Thaxter's words. When I focus on how good it feels, I forget how little I might know. (And you really need to only know a little, just a little, to have a cutting garden!)

Let this book remind you what you likely already know, what you were likely born knowing, that the simplicity of growing a flower and being outdoors provides immense joy. Be encouraged to use your garden as a place to fall apart and put yourself back together, even if your garden is just an idea in your head or a sketch on a piece of paper. Let it be a place to tap into the wonder that is discovering a ladybug or a bee asleep in a flower petal. Let the flowers you grow and give to others remind them to be here in this beautiful space, if only for a moment.

How to Use This Book

It is my hope that you use this book as inspiration, and as a friendly guide to get yourself growing flowers and increasing joy and happiness in your life. There are many gardening books that can explain the techniques and the science behind growing flowers, and you'll likely turn to many of those when the time comes. But for now, let's do this together, season by season, as if we were in a supportive garden club. Let me help you walk through your first year.

Winter
Arrive & Inspire

> When the snow is still blowing against the window-pane in January and February, and the wild winds are howling without, what pleasure it is to plan for summer that is to be.
>
> —Celia Thaxter

To begin a garden, you must first dream it to life. This dreaming state of gardening is what the writer Katharine S. White called, in her book *Onward and Upward in the Garden,* the garden of the mind's eye. A garden does not begin in the soil, though we will discuss the virtues of good soil later; it starts in the soul of its gardener. It is a story we tell by tending to the earth, day after day, until the frost comes.

For many of us, this story begins with people who came before us. I often wonder if everyone's garden is a love letter to their grandmother or a forgotten space in time. A garden is also an incredible source of healing. Once humans gained the ability to stay in place through the cultivation of seeds (and the creation of the fence to keep browsing critters out), we looked to the garden for nourishment and medicine.

Your garden will hold room for failure and function as an expectation-less space to roam. Nothing will ever be just as it is supposed to be. Things will surprise you. The garden will never be done, complete, or as you imagined it. Often, it will present itself as a living reminder of all your misgivings (too much water, too little compost, not enough sun). But by some grace, it will be beautiful all the same.

So let us start by remembering where we came from. After all, isn't that where all good stories begin?

A Flower Is a Memory

Consider, for a moment, some of your earliest memories of flowers. Likely it's not that difficult, as flowers and gardens create an intense sensory experience. If I remind you of a lilac just now, can you close your eyes and almost taste the smell in your mouth? Can you recall the vibrant shade of yellow glowing beneath your brother's chin while you held a buttercup at just the right angle? How about a long-forgotten hiding spot tucked beneath boxwood branches that held you in a place of safety when the world felt oh so big?

How do you feel when you call these memories to mind? How meaningful would it be if you could create a garden that pulled these warm, seemingly faraway places and people to the front of your mind every time you stepped out the back door?

Creating a cutting garden that reflects where you have been, and perhaps where you want to be, allows one to move through the growing season with purpose. It will act as a force of motivation when the weeds grow beyond your ankles, the groundhogs eat your seedlings, and the sweat rolls down your tired back. This garden of the mind's eye will anchor you to the soil beneath your feet and the seed in the palm of your hand.

So let us begin this year slowly, by dreaming of flowers in vibrant colors and shapes. Later in winter, as the days become longer, we will narrow our vision to create a simple, easy, and productive garden.

Finding Inspiration

They say imitation is the highest form of flattery, and if that is true, then I hope Celia Thaxter interprets each flower I plant as a running list of accolades. Of course, Celia hasn't been of this living world for over a hundred years, but her journals and sketches of the island garden she tended are a constant source of inspiration for me. I turn to them often. But this doesn't mean that I use them as a technical point of reference. They are there as a touchstone, and that's entirely different. What is your touchstone? Is it

a garden you remember as a child, or one that you pass by on your walk to work? Is it a public garden you've visited, or a garden from a movie? Is it the landscape of a place you've always loved to visit? Think about the things, whether they are books, films, memories, or moments, that you turn to in your mind again and again.

Thaxter's garden was, in my opinion, nearly impossible to cultivate as it required that she board a steamship each spring, leaving her life on the mainland to spend summer on the island, loading the ship with crates full of sand containing small eggshells of seedlings she had started earlier in the year. On the island she experienced volatile storms and less-than-ideal soil conditions. It's not like she had it made when it comes to "perfect weather," which, of course, doesn't exist, especially these days. Her garden, filled with phlox, pinks, and hollyhocks, showcased her dedication to turning a weather-beaten, cold island into a stunning cut flower garden that will forever remind me that the impossible is possible.

It's essential to identify people in the gardening world who excite you. Discovering gardeners who live in similar growing areas or gardening zones can be incredibly useful. While I admire and pine over lots of beautiful UK-based gardens designed by renowned horticulturalists, the truth is the harsh cold in Maine makes some varieties of blooms nearly impossible to grow where I am. This takes a little bit of research, but if you're dreaming up a garden, you're likely already doing it intuitively.

Find Your Inspiration

» Discover your growing zone with a quick internet search. This information will help you determine your average first and last frost date. With these dates in mind, you will be able to select which flowers you can grow in your garden and know when to plant them.

» Look for gardeners in a similar climate to your own, and take note of the flowers they are able to cultivate.

» Create a photo album on your phone or computer entitled Garden Inspiration. If you, like me, have scattered images of flowers taken while visiting a friend's garden or walking in the woods, consider organizing these images. Slide photos into your garden inspiration album so they are easy to call up from time to time. Take note of any flowers that stir something inside you and ask why. Is it the shape, the color, a memory, perhaps? Jot these traits down. When it comes to working with digital content like this, I encourage you to put a timer on your phone for fifteen minutes. When the timer goes off, take a break. Consider printing these images and hanging them on a wall or inside your closet door.

» This leads to my next suggestion of enjoying floral images in books and magazines. There is no time limit here—flip pages until your heart is content. Don't forget to leave an inspiring page wide open on the kitchen table so you are reminded of the beauty of flowers each time you walk by.

My Beloved Garden Books

» *The Land Gardeners: Cut Flowers* by Bridget Elworthy and Henrietta Courtauld

» *Cultivated: The Elements of Floral Style* by Christin Geall

» *On Flowers: Lessons from an Accidental Florist* by Amy Merrick

» *From Seed to Bloom: A Year of Growing and Designing with Seasonal Flowers* by Milli Proust

» *Everlastings: How to Grow, Harvest, and Create with Dried Flowers* by Bex Partridge

» *Onward and Upward in the Garden* by Katharine S. White

» *Floret Farm's Cut Flower Garden: Grow, Harvest, and Arrange Stunning Seasonal Blooms* by Erin Benzakein with Julie Chai

» *Green Thoughts: A Writer in the Garden* by Eleanor Perenyi

» *An Island Garden* by Celia Thaxter

Explore Your Growing Space

Put on your warmest winter socks, boots, and gloves, go outside, and observe your growing space. Feel free to bring a sketchbook or notebook with you. While outside:

- » **Get your bearings.** If you aren't sure how to identify north, south, east, and west in the garden, use the compass on your cell phone. Flowers that are south-facing get the most sun. Over the next few days, observe the sun's arc over your growing space. Since it is winter, the days will be shorter, but you can develop an idea of how light travels in your growing space and where shadows are cast. Begin to consider where you want your cut flower garden to take shape.

- » **Find a water source.** Identify outside spigots, determine how far away they are from your potential garden, and jot down what supplies may be needed. Think hose length, watering cans, or perhaps irrigation systems.

- » **Complete a simple landscape intake.** While walking the yard, take notice of any shrubs, flowering trees, or previous perennial gardens that exist in your landscape. When you return inside, research or recall these items' approximate bloom window and color. Working with what you already have in your landscape can help you create beautiful floral arrangements with little effort.

- » **Sketch your space.** Using a tape measure, determine the length and width of your growing space. Even if it is a small back patio or a few simple containers, a basic sketch of the space will help when selecting flowers. This does not need to be anything fancy! Just an outline. We will add more detailed information later.

I encourage you to take your time with this exercise. You may want to explore your growing space at different times of the day. So many of us rarely feel fresh air on our skin in winter. It can be incredibly grounding to pause and move slowly. Don't forget to keep your eyes wide open for trees or shrubs beginning to show buds. Any small sign of spring can lift your winter-weary heart.

Explore Your Design Aesthetic

Before selecting which blooms you will grow this year, let us narrow our view and study bouquets or floral arrangements that speak to our personal design aesthetic. Working backward by visualizing arrangements before selecting blooms can help you identify the types of flowers you want to grow and how many seeds you should sow this spring. As you review, ask yourself:

» **What ratio of bold focal flowers to filler or foliage am I drawn to?**
For example, I like floral designs that accentuate all the drama in a long, arching single stem. Big gaps and empty spaces in an arrangement give me time to ponder each individual bloom; thus, I don't grow many filler flowers or foliage. If you enjoy a full bouquet with less focus on individual blooms, add lots of filler flowers like yarrow or feverfew to your garden.

» **Do I like a more structured or a wild style?** If it's the latter, consider textured blooms like nigella and cosmos. Or, if structured, flowers with clean lines should interest you. Snapdragons and foxgloves come to mind.

» **Will my arrangements be wrapped and hand-tied for gifts?** Or will I be arranging over and over again in a beloved vase? Also, consider any special events, such as birthdays or celebratory dinners, during your growing season. Planting flowers that remind you of the people you adore and then sharing them with that person is a beautiful practice. Consider what flowers remind you of a beloved person and aim for that bloom to pop just before their special day.

Take a few moments now to sketch or save images of bouquets you find visually appealing. While I don't consider myself an artist of any sort, sitting down with a crisp watercolor notebook and paint is incredibly peaceful. Sketching and painting the flowers you plan to grow, while admiring them in a photo, helps you begin to understand the shape and subtle color tones of each bloom. Don't worry about how "good" the paintings are. Think of them as part of your garden inspiration and remove expectations about their value.

Flowers for Cutting

The purpose of a cut flower garden is to grow blooms meant to be harvested, arranged, and brought indoors. This is why it is necessary to ensure you have different types of flowers in a cohesive color palette that bloom at the same time. Color palette is an extremely important decision when planning a cut flower garden. Without a clear plan, you might find you have armfuls of blooms that feel conflicted when put together. Color preference is also incredibly personal, so take your time with this decision.

Types of Flowers

When we create a cut flower garden, it is important to cultivate blossoms with a variety of shapes, textures, and heights. This allows us to create bouquets and arrangements that are visually interesting. The four types I grow in my garden are called focal, secondary, and filler flowers, and foliage (or greenery).

» **Focal:** Also known as mass flowers, these tend to be the visual lead actors of every bouquet. Your eyes are immediately drawn to these blooms' lush and expansive nature. Think large decadent tulips, dahlias, and roses, or uniquely colored sunflowers.

» **Secondary:** Often, this is a flower that has an interesting shape but is less flashy than a focal flower. It adds dimension or line to an arrangement. Foxgloves, delphiniums, and snapdragons are fine examples of line-shaped flowers that work well in a bouquet. Also consider smaller zinnias or cosmos as secondary flowers.

» **Filler:** These blooms are just as they sound—something to soften the hard lines and fill the gaps between your focal and secondary flowers. Often they have multiple small blooms on a single stem. My favorites include yarrow, Queen Anne's lace, feverfew, and flowering herbs.

» **Foliage:** I only add foliage to bouquets or arrangements when I see greenery that shouts to me from across the garden. Wild raspberry branches, boxwood, and ornamental grasses are all beautiful in their own right. Personally, I dislike when foliage is used to fill in or overstuff a bouquet.

Considering a Color Palette

When growing flowers to sell, I would often look at wedding color trends for the upcoming season and attempt to mimic these out in the field. As I grow more for myself and less for others, though, I'm drawn to color palettes from works of art that inspire me or from natural color patterns that I discover outdoors. A bearded iris, for example, might be primarily deep blue purple but have small, almost paintbrush-like details of white, black, and yellow. These colors speak to me! But there are so many color palettes interwoven in different bearded irises. If you have a handful of must-have flowers for your cutting garden, look closely at images of these flowers for complementary color combinations right in their petals. If you were using the iris I described for color guidance, you might consider growing white snapdragons, pale yellow 'Lemonade' cosmos, white yarrow, delicate purple echinacea, and vibrant yellow-petaled rudbeckia with its black seed center.

Of course, if this feels a bit too much, a neutral palette is by far the easiest and will give you a wide variety of flower shapes and sizes to choose from. If you select neutral-colored flowers, you might consider adding a single contrast color to make things pop. Most often I use orange or deep purple to contrast a more neutral palette. These two bold colors can be found from spring to fall in various flower shapes. The most important thing to consider is that too much can be too much. Staying simple is always the best place to start. When beginning a cutting garden, it's good to remember that we can't grow *all* the flowers in the world, and we have to choose. It's the first exercise in trusting your gut and figuring out what your style is.

A Mix of Flower Varieties

A good cutting garden should have a mix of annual flowers (flowers that only last one season and need to be replanted each year) and perennials, which are planted once and return on their own year after year. There is also a third category called biennial, which are flowers that return every *other* year, like foxgloves and hollyhocks.

Perennials often fill the garden space in early spring before your annual seedlings produce blooms that can be harvested. Although perennials can be costly at the onset, they require less maintenance and watering once established. I highly recommend adding hardy native perennials appropriate for your environment, as they support pollinators and are better equipped for fluctuations in climate.

My Favorite Perennials

I had the good fortune of moving into a home with an already-established perennial garden. A handful of the flower varieties were familiar to me, such as peonies, roses, and lilies. But many of these blooms I felt clueless about, and differentiating between a weed and a flower in early spring seemed impossible. I spent mornings wandering the garden and identifying plants, marking their common and proper Latin botanical name down in a tiny notebook. Mostly I observed for the first few years and let the well-thought-out garden speak to me in the waves of color that flowed harmoniously through the growing season from May to October with ease. In winter, I'd open my notebook and recall each unknown plant while saying the proper names out loud, pronouncing them poorly, and jotting down growing instructions. In retrospect, I was trying a bit too hard. A perennial garden mostly just needs to be enjoyed and supervised once established. Simply weed and water here and there, and let nature do its thing.

Once I let go a bit, I decided to add new perennials that reflected my personal style. Inspired by a visit to Piet Oudolf's perennial double border at the New York Botanical Garden, I added and experimented with grasses and bulbs. This has allowed me to slowly add lines and interest to the garden in winter, while leaving most of the established blooms in place.

Through this carefree exploration of perennial blooms, I've identified easy-to-grow plants that work well as cut flowers. Here is a list of these varieties, which vary in bloom window, shape, and color, ensuring you have lovely fresh flowers all season long.

Astrantia major

This beautiful star-shaped flower, with its crisp, structural petals, tends to be a conversation starter as many gardeners are unfamiliar with it. I enjoy harvesting a few stems and leaving the rest on the plant to dry and fade over the course of the season. Even when this bloom is not living, the petals tend to hold in place and add interest to a garden bed or your vase. While this flower appears delicate, it is quite hardy. Another benefit is that it comes in various shades of white, pink, and purple, which means it will work in many color palettes. It is wonderful as a dried flower, and my favorite variety for this purpose is 'Abbey Road', as it holds a deep maroon color. Plant it in a spot that receives morning sun and afternoon shade; it can handle wet soil.

Black-eyed Susan (*Rudbeckia* spp.)

I must admit that my favorite variety is 'Caramel Mix', with its stunning deep maroon centers and golden-hued petals. However, rudbeckia's most common variety is recognizable by its bright vibrant yellow petals and large black center. It may have been the very first wildflower I ever encountered, as it pops up along sunny stretches of country roads all over New England. In truth, it seemed such a common flower that I wasn't initially interested in formally growing it in my garden. But I've come to add many varieties of rudbeckia over the years, as they provide reliable color to a garden in late season. Though not as prolific of a producer as cosmos or zinnias, this plant will yield many stems over the course of the season. Plant in well-drained soil in full sun.

Blanketflower (*Gaillardia* spp.)

A vibrant red, orange, and yellow single-petaled bloom that is native to North and South America. This daisy-like bloom works well as a fresh cut or pressed flower. I also enjoy using the seed pod center in a bouquet after the petals have fallen off. This perennial will expand rapidly over the years and puts on bright color from early summer to late autumn. Plant in full sun; it is known to tolerate poor soil.

Coneflower (*Echinacea* spp.)

I adore the upright strength of this native bloom. It seems to slowly pull itself from the earth on an impossibly stiff, textured stem and then send its colorful petals with force back toward the earth. It stands out in a bouquet, thanks to its unique petal shape. Coneflower comes in a great variety of cultivars with different styles and colors. However, the true native is by far my favorite, which has a large seed head and petals that just barely touch. Plant in sun; it will tolerate rocky soil but does not like to be wet.

Delphinium spp.

There is a lovely little red house at a curve in the road on the way to the coast here in town. Each year I admire the incredible delphinium collection at the east end of the yard. Vibrant blue petals in small star formations climb the stems, which are staked and held in place with twine. As the years have carried on, the collection has swelled. It appears a stoic army, in their most stunning royal blue attire, holding strong toward the sun. This flower can be grown in shades of blue, purple, red, pink, and white. Plant in full sun, and stake for extra protection if planted in an area with wind.

Foxglove (*Digitalis* spp.)

Visually stunning, this tall line flower is delicately decorated in speckled bell-shaped blossoms. Some believe it earned its common name because it appears that the petals might slip onto a fox's paw and function as a tiny colorful glove. Much like coneflower, I'm intrigued by the way the petals seem to melt back down toward the earth. Historically, foxglove was used for its medicinal properties, producing digitoxin, which can excite the muscles of the heart through an increase of calcium. However, foxgloves are also extremely poisonous and toxic. Some believe Vincent van Gogh was being treated with foxglove during his yellow period, as visual effects from foxglove poisoning include halos around light, similar to those seen in *The Starry Night*. So, take care when handling this bloom around pets and children, and be sure to wash your hands thoroughly after working with it. A woodland plant, foxgloves prefer shade and rich, organic well-drained soil.

Globe Thistle (*Echinops* spp.)

I can't tell you how many times I've poked myself on globe thistle. It is easy to forget that this bloom is covered in tiny spikes mostly because from afar it looks to be a whimsical lollipop floating in your garden row. Consider wearing garden gloves when harvesting! This plant adds lovely texture to bouquets. Grow in dry or sandy soil in full sun. Do not plant in moist soil, as the root is prone to rot.

Goldenrod (*Solidago* spp.)

While many people might view this as a common roadside flower, the color of goldenrod absolutely thrills me. It looks especially decadent as the sun sets, catching the yellow blossoms and turning them to gold as they cast a long shadow to the ground below. At first, I simply cut intermittently from the wild goldenrod growing at the edge of the yard. But in recent years I've added it to the back row of my cutting garden (as it can grow up to six feet tall) and left the wild solidago for the pollinators. This flower is multipurpose: use it in arrangements, as a dried flower, or for making natural dye. Plant in dry, sandy soil in the sun to part shade.

Hollyhock (*Alcea rosea*)

One could argue that this bloom is most often associated with a cottage-style garden. Hollyhock can grow to incredible heights, making even an adult feel dwarfed and childlike in its company. The large, lush blooms open from the bottom of the stem to the top, so be sure to look to the ground first while you wait in anticipation for the petals to expand. Consider planting it in front of a solid surface, such as the side of a house or a shed. This allows the expansive shape to visually pop. Hollyhock comes in a variety of colors and can grow three to six feet tall. Be sure to place them in the back row of your cut flower garden, so they do not shade out the rest of the flowers. Plant in well-drained soil and full sun and consider staking to protect them from wind.

Iris spp.

Perhaps you, like me, fell in love with irises by way of van Gogh. I was lucky enough to find our perennial garden swelling with well-established irises, slowly climbing over the informal border to the yard at large. The days leading up to an iris finally dropping its outer petals to expose a delicately ruffled interior are full of excitement. In the past few years I have expanded my iris collection beyond the traditional purple. These flowers will grow quickly and need to be divided every three to five years in late summer. Plant in full sun; be sure the soil is well draining as the rhizome can rot.

Lady's-Mantle (*Alchemilla mollis*)

Lady's-mantle creeps low in the garden beginning in early spring. The wide and softly textured leaves are something I always pause to touch when walking the garden each morning. In early spring the cupped green leaves seem to support a collection of fleeting delicate yellow-green blossoms. The bloom window of this flower is similar to that of peonies, and the two complement each other beautifully in a bouquet. It can be planted in sun to part shade, and tolerates most soil types.

Lavender (*Lavandula* spp.)

Please plant lavender. It is such a gift to cut a fresh sprig and share it with someone. If you share it with a child, instinctually they will close their eyes and take a long, thoughtful breath with a big exhale. Remind any adults who forget this to do so. When the lavender blooms in my garden, I enjoy wandering back and forth, running my fingertips slowly across each flower. That is, until the bees arrive, at which time I simply stand back and listen to their wings beat in tandem. The intense hum that fills the air can slowly pull me from my ever-busy mind. I grow the variety *Lavandula* ×*intermedia* Phenomenal as it is cold hardy. You can count on these lovely scented purple flowers from late spring through fall. Plant in well-drained soil in full sun.

Peony (*Paeonia* spp.)

The petals of a peony are soft, lush, and romantically fragrant. Peonies feel full luxury, with flower heads so impossibly large they cause the flower stems they sit upon to dip and bend down to the earth. This flower is so beloved that almost every gardener I know has a peony or two tucked into their landscape. When they bloom in late spring, most kitchen tables have a small bundle in a vase, a sure sign that summer has nearly arrived. Each year I make efforts to add more peony varieties to my garden, expanding my collection beyond the traditional pink to include shades of deep maroon, red, and white. While the blooms are often short-lived stunners, the wide leaves on each stem can be used as a bit of greenery in your bouquets all season long. Be mindful not to trim too many leaves, especially from a young plant; the peony needs the leaves for photosynthesis. Plant in full sun, with well-drained soil.

Pincushion Flower (*Scabiosa* spp.)

Pincushion flower has a lovely shape: a wide, circular center adorned with a single delicate wrinkled-looking row of petals around its exterior. It also offers a never-ending parade of colors, from moody maroon to crisp white and delicate periwinkle blue. All of this beauty sits atop a thin stem that bends in various directions, making it interesting all on its own in a bud vase. It also acts as a lovely secondary flower in bouquets or arrangements. This flower will bloom from June through October when cut regularly. However, consider letting lots of the blooms go to seed and use the pods in dried flower arrangements. Plant in well-drained soil in direct sunlight. A known friend to pollinators.

Solomon's Seal (*Polygonatum* spp.)

I enjoy this bloom (*above, left*) all season long as it is constantly changing. In late spring the wide, smooth leaves are adorned with small white floral bells. As the season progresses the flower bits fade, and by autumn the leaves have shifted from green to a yellow orange. The natural transition of color and shape seems to complement other garden flowers that come in and out of bloom. In early spring Solomon's seal will add a lovely bit of foliage for pink peonies, and the yellowed leaves act as a delicate support for dahlias in autumn. This plant enjoys full to part shade and well-drained soil.

Tickseed (*Coreopsis* spp.)

This bloom (*above, center*) can fill large areas of a cutting garden and creates a cloudlike array of color that maintains itself with little effort for most of the growing season. I enjoy tucking it in the corner of a row of flowers. A wonderful native perennial, coreopsis has incredible vase life, and comes in a range of colors. *Coreopsis* 'Red Satin' and *C. verticillata* 'Moonbeam' are two of my favorite varieties. Also consider planting golden tickseed, also known as dyer's coreopsis (*C. tinctoria*), and harvest the petals for natural dyes. Plant in full sun, in well-drained soil.

Yarrow (*Achillea* spp.)

Yarrow (*above, right*) has a long, sturdy stem adorned with a flat-topped collection of tiny blossoms. This flower feels very rooted and grounded compared to the more whimsical flowers in the garden that bend and bow in the breeze. Use this bloom to create blocks of color in a bouquet or dry it for everlasting flower arrangements in the winter months. This incredibly rugged perennial can take extremes, from drought to buckets of rain. Yarrow will produce flowers the first year from seed, but really hits its stride in year three. It tolerates poor soil but enjoys the sun.

My Favorite Annuals

I began my cut flower garden solely with annuals for a few reasons. First, the seeds were incredibly inexpensive, making it the most economical way to begin a garden. Second, annuals can reliably grow flowers from a single seed all season long and, as my garden was initially quite small, this was incredibly important.

Ageratum spp.

If you are hoping to add a bit of blue to your garden, ageratum is a perfect flower to do just that. Its color stands out not only in the garden, but in a bouquet, especially when paired with neutral or vibrant blooms. While walking the garden in the evening, ageratum in a row seems to glow just before the sun goes down. The blue blossoms are only overshadowed by the textural leaves that line the stem. These leaves are wide, ridged, and rippled, making them lovely in a vase arrangement. After a hard cut to the ground in early summer, a second flush of blooms arrives in autumn. This bloom loves full sun but can also work in part shade.

Calendula officinalis

I enjoy growing this flower in the most vibrant orange variety one can find. In a row, or when planted in a small pot or at the edge of a garden, the wide-open, petaled face of this bloom is decidedly cheerful. When using it in a bouquet, its delicate daisy-like shape provides a bright pop of color without stealing the show. When harvesting calendula, a sticky residue will often run in between my fingers, causing me to pause and consider its long history as a medicinal plant. This flower can tolerate full sun to part shade and likes well-drained soil. I find it to be a reliable self-sower, dropping seeds in the soil before winter and blooming the following spring.

Cockscomb (*Celosia* spp.)

Cockscomb is easy to identify from a distance in one's garden. Its unique narrow, feathery plumes spend the entire summer slowly growing and reaching toward the heavens. This flower is often unperturbed by neglect or wild extremes in weather, making it a reliable bloomer for even the most absent-minded gardener. 'Flamingo Feather' is my favorite variety, which comes in a lovely mellow tone of pink that gently antiques when dried. This bloom adds a pop of interest to a bouquet or can be used in winter wreaths after drying. As these blooms are native to the Mediterranean, they are a star producer in the high heat of summer. It is happy in full sun and well-drained soil.

Cosmos spp.

These flowers seem to visually elevate a simple garden to one that you might find in a fairy tale, with long, delicate stems that bend in the wind and rain, and soft, wiry bits of greenery that stretch to impossible heights by season's end. Each flower is topped with delicate petals in shades of white, cranberry, pink, or yellow. Cosmos represents to me the beauty of a hyperlocal bouquet. They are impossible to buy at a grocery store as they do not travel well. While the blooms only last for a few days in a vase, if you make sure to cut some stems when the buds are just about to open, they should last longer. The Double Click variety, in white or cranberry, is playful. Plant in well-drained soil in full sun.

Dahlia spp.

A late-season stunner, dahlias come in just about every size and color imaginable. Perhaps the most well-known is the dinner plate–size 'Café au Lait', which functions as an incredible focal flower, especially in bridal bouquets. These flowers start slowly, putting on luscious deep green leaves for months, until finally one day a single stem emerges with a large obvious bud. The scent of a dahlia flower when cut is equal parts dirt and sweet nectar. Some varieties have petals that curl up onto themselves in plentiful rows, while others have a single row of petals with rounded or pointed edges. The never-ending array of colors, sizes, and shapes that dahlias offer means that you could grow just this one flower and be rewarded with fantastic diverse bouquets. Plant their tubers, which look like strange-shaped potatoes, in the soil when the lilacs bloom. Dahlias like full sun and rich, well-drained soil.

Globe Amaranth (*Gomphrena* spp.)

This flower produces small pom-pom–like orbs that add visual interest to a bouquet and works well as a dried flower. As it is quick to germinate and doesn't take up much space, I sow globe amaranth seeds all season long by tucking them into bits of vacant soil. Nearly every row in my garden has gomphrena poking out and bending around well-established perennials. Enjoys full sun and thrives in most soils.

Hyacinth Bean (*Lablab purpureus*)

I planted the variety 'Ruby Moon' when researching ways to enhance my soil naturally. A tried-and-true method for adding nitrogen back into the soil is to plant legumes. And as I'm always looking for things to grow that can be used as cut flowers, 'Ruby Moon' hyacinth bean was an ideal addition to my garden. This legume just happens to be an incredible climber that produces delicate light purple flowers and deep purple bean pods. I enjoy encouraging it to climb up repurposed sticks tied together with twine, which I anchor into the end of my garden rows. Plant in full sun and well-drained soil.

Lisianthus (*Eustoma* spp.)

I began growing this flower many years ago when roses were hard to come by in the wedding world. Florists were clamoring for a good substitute, and lisianthus fit the bill. This gorgeous flower is native to the United States in places like Nebraska, and is also known as a prairie gentian. Lisianthus comes on very slowly, taking roughly six months to bloom from seed, so consider purchasing it from your local nursery. As they are slow bloomers, they will absorb precious garden space for many months, which is not ideal in a small cut flower garden or container. But even just a handful of these flowers, planted tightly with other long-stemmed blooms, can be enough. They ever so slowly put on greenery in very early spring when the earth is cold, and then seemingly overnight the thin-yet-sturdy stems shoot upward, adorned with luscious petals. Plant in rich soil, in full sun.

Nigella spp.

A truly unique bloom, it is often the flower that stops people in their tracks when we wander the garden. Star-shaped blossoms of blue, pink, and white with gentle fernlike foliage ensure a poetic, earthy element in every bouquet graced with its presence. Also, consider enjoying it in the garden, letting the petals drop and harvesting its seed pods for dried arrangements. This is a wonderful bloom to direct sow, and when I can remember, I direct sow these seeds two to three weeks apart in the garden. Plant in full sun and well-drained soil.

Panicum capillare 'Frosted Explosion'

This grass can look a bit unruly in a cut flower garden; it is a wildly textured burst of foliage that shimmers in the morning sun, hence its name. I tend to plant it on the edge of my garden or in forgotten bits of dirt along the driveway. While it can look lovely in a bouquet, this is my go-to item to dry and put in seasonal wreaths. Enjoys full sun and well-drained soil.

Snapdragon (*Antirrhinum* spp.)

It is unimaginable that a tiny snapdragon seed can transform in just a few short months into an elegant, delicate-petaled line flower. The snapdragon adds class and distinction to your garden, seemingly anchoring wilder and more whimsical blooms like cosmos and nigella. I plant rows of snapdragons in my garden as they are deer-resistant and can be planted in early spring. If the winter is mild, often these blooms will self-sow, producing flowers the following year on their own. Plant in well-drained soil in bright sun.

Strawflower (*Xerochrysum bracteatum*)

This flower comes in a wide range of colors, from moody auburn to light pastels. Strawflower presents almost as a dried flower when in full bloom. Hard, pointed petals open to reveal a soft seed center. This flower will slowly open its petals in the morning sunlight and by evening pull them back in to cover its precious seeds. It grows tall in the garden, three to four feet, and thus should be planted in a back row. It enjoys full sun and well-drained soil.

Tassel Flower (*Emilia sonchifolia* var. *javanica* 'Irish Poet')

This lesser-known flower was something I stumbled upon in a seed catalog and thought, why have I never heard of this precious little bloom? I'm so very glad I took a chance and decided to grow it, because it became the flower that florists requested most often in my first few seasons growing to market. This bloom is incredibly easy to grow and will put on new flowers as you cut it, although it is not as prolific of a producer as, say, zinnias or cosmos. It has a very long, slender stem that can grow up to two feet, and is topped with a tiny little bright orange tassel-like blossom. Sow in well-drained soil in a place that receives full sun.

Zinnia spp.

The happy-go-lucky, easy-to-grow, never-ending bloomer of a cut flower garden, zinnias come in an overwhelming number of shapes and sizes. In my garden I dedicate an entire row to these flowers, and pack them in tightly. A single overflowing row of these beauties gives the illusion that you are a well-seasoned expert gardener, even if it's your very first year sowing a seed. I tend to grow smaller zinnias that function as secondary flowers. These include *Zinnia haageana* Jazzy Mixed, *Z. elegans* 'Zinderella Peach', and *Z.* Oklahoma Series. Plant in well-drained soil in direct sunlight.

Herbs and Other Edible Elements

While the majority of what I grow is flowers, there are many ways to enhance your garden and your arrangements with edible elements. Raspberry branches, for example, can be an excellent addition to a bouquet. A bouquet scented with fresh herbs is also unique to a home garden, where your flowers and herbs might even be intermingled in the same space. However, the vase life and harvesting timeline of herbs can be tricky, which is worth keeping in mind.

Basil (*Ocimum* spp.)

My favorite variety is 'Dark Opal', which is bold and rich in color. Do not harvest this basil until the stem has become hard, almost like a woody shrub, and it starts to flower. All basil is prone to drooping if you cut it before this stage or harvest it during the heat of the day.

Bouquet Dill (*Anethum graveolens*)

This herb looks lovely interplanted among cosmos, a tip I learned from the Land Gardeners, Henrietta Courtauld and Bridget Elworthy. It provides a wildflower aesthetic in what could be perceived as a simple straight row. The flowers of this dill can become enormous, so it is wonderful included in a large vase to make a statement.

Feverfew (*Parthenium* spp.)

Technically an herb, though I think of it as a flower, these cheerful tiny white blooms should be planted in succession to ensure you have them on hand all season. It is a wonderful filler flower in bouquets.

Mint

Mint adds incredible texture to a bouquet but be careful where you place it in your garden as it spreads with passion! It is best to put it in a pot or raised bed to contain it. Apple mint (*Mentha suaveolens*) is a lovely variety to grow in a cut flower garden as the stems are incredibly long and it puts out beautiful purple blossoms. I also have enjoyed working with native mountain mint (*Pycnanthemum* spp.) in recent years; it is maintenance-free in the garden and stands out in a bouquet, thanks to its unique shape.

Fruit

When it comes to fruit, I'm all about simplicity, so easy-to-grow berries are a must. A raspberry warmed by the sun is one of life's pleasures. In high summer, we spend most mornings and evenings sneaking to the raspberry bush, hoping to gobble handfuls of berries straight off the branches. I use raspberry leaves in my bouquets all season long. In high summer, people seem to adore a surprise raspberry in a wrapped bouquet; in fall, the leaves begin to yellow, reflecting the return to autumn. There are thornless varieties of raspberry bushes, so keep this in mind if you don't want to be tasked with constantly reminding people to watch their fingers when they handle your flowers.

As we live in Maine, we also have blueberries around the garden. I recommend planting two or more different varieties that bloom at the same time. This will encourage cross-pollination and increase your yield. Finally, I love to harvest blackberries early while they are still green. Once they have ripened, they make incredible natural dyes.

EDIBLE FLOWERS

Many varieties of flowers can be used in the kitchen. Nasturtium, for example, which produces bright orange and red blooms, is perhaps the most well-known edible flower. Calendula, roses, lavender, French marigolds, and lilacs all fall into the edible category. Use them to make a simple syrup for cocktails, to dress up a salad, or to create a decadent pavlova at the meal's end.

Sourcing Flowers

The very first garden I tended was what you might call a postage-stamp garden. The space measured roughly four by eight feet, and each spring after the final frost we would load up the children and venture to a nearby greenhouse. Once there, as the children ran the aisles and knocked over pots, I'd slowly roam and pick up flowers of various heights and colors, visualizing where I might squeeze each plant into our tiny plot. Back at home, we would dig into the soil, pack everything in tight, give it a good soak with the hose, and, lo and behold, we had a garden!

If you are new to gardening and the idea of seed starting seems overwhelming, let me assure you that sourcing flowers from a reputable greenhouse is a fantastic

Narrow Your Flower Selections for This Year's Garden

Consider taking an afternoon to stroll through a nearby art museum and observe your reaction to various works. It is surprising how quickly patterns will emerge and what colors and textures hold your attention. Or take a walk in the woods or by the sea, observing colors and forms that speak to the places in the physical world you love. Determine your cut flower garden color palette.

Write down one or two flowers in each floral-type category (focal, secondary, filler, and foliage) to narrow the vision of the garden of your mind's eye. This can be difficult as there is an endless list of stunning blooms for cut flowers. Remember, your gardening journey is long. Start small to ensure you understand the different personality of each flower.

way to begin. However, don't rush the process! A good greenhouse will avoid putting out tender annuals like zinnias or cosmos until they can be safely planted in the soil. This means you may have to wait a week or two after your last frost date to start digging. Also, keep your eyes out for garden-club plant sales or local farmers selling seedlings. These are both fantastic ways to support local businesses.

Starting Your Flowers from Seed

For those of you on the second or third year of your gardening journey, you might be feeling ready to start some flowers from seed. This takes a bit more planning and attention to the calendar and logistics, which is why it may not be right for your first year. And that is totally fine. But if you're ready, I will now describe how to create a seed-starting schedule and space for growing seeds. Even if you feel this is not your year to start from seed, go ahead and read on. It is highly likely seeds will pique your interest at some point. And once you start flowers from seed, you'll never go back!

> Take a Poppy seed, for instance: it lies in your palm, the merest atom of matter, hardly visible, a speck, a pin's point in bulk, but within it is imprisoned a spirit of beauty ineffable, which will break its bonds and emerge from the dark ground and blossom in a splendor so dazzling as to baffle all powers of description.
>
> —Celia Thaxter

Flowers That Can Be Direct Sown

If you are desperate to put your hands in the dirt and plant some seeds, consider the following flowers that can be direct sown into the earth in early spring. These are plants that require no grow lights, seed-blocking kits, or watering in March.

I want to draw attention to the phrase "direct seed as soon as the soil can be worked," which is often a directive on a seed packet and yet can be a bit confusing to the beginner gardener. What does "ready to be worked" mean, you might wonder? You will know your soil is ready to be worked when the top few inches are dry and crumbly. For those of you with clay soil, like myself, this will take a few weeks longer than those with sandy soil. However, if you are working in a raised bed or container, it is possible to sow seeds earlier as you are not working with previously frozen soil. The following is a list of some of the flowers I'm able to seed directly into the earth in very early spring, which is always quite an exciting moment!

CALENDULA OFFICINALIS

Direct sow as soon as the soil can be worked. This flower is not picky when it comes to soil type, and it can tolerate part shade. Plant seeds roughly five inches apart. Calendula germinates quickly, so it is a wonderful bloom to plant with children, who tend to be impatient gardeners at the start.

CHINESE FORGET-ME-NOT (CYNOGLOSSUM AMABILE)

These delicate small blue flowers have a long stem. Direct seed as soon as the soil can be worked.

LARKSPUR (CONSOLIDA SPP.)

Blue and white larkspur are stunning, and a wonderful line flower to dress up the garden or a simple summer bouquet. Direct sow two to three weeks before the last frost.

LOVE IN A MIST (NIGELLA DAMASCENA)

Direct sow when soil temperatures have reached 60 degrees, roughly two weeks before the last frost. Continue to sow every three weeks for blooms all season long.

SHIRLEY POPPY (PAPAVER RHOEAS)

Known for their bright red petals and tall stems, poppies are a cottage-garden classic. Although their vase life isn't the best, having them in your home for even a day or two by the bedside table is heaven. Direct sow one month before the last first date.

SILVER TIP (*TRITICOSECALE* SPP.)

One of my favorite ornamental grasses, which I never seem to have enough of, with tall, sturdy stems and, as the name suggests, a distinct silvery glow. Direct sow one month before the last frost date.

SWEET PEA (*LATHYRUS ODORATUS*)

Oh, sweet pea, how I loathe and love you. I must admit I have yet to master sweet peas, mostly because just when they seem to start blooming, the weather becomes too hot and suddenly they tire. However, I keep trying! You can transplant these out when the ground can be worked, or when the daffodils start to poke through. As long as your nights don't dip below 20 degrees, it is safe to direct sow.

Succession Planting for Annuals

I am often asked: How do you still have [insert beautiful flower here] in August? Didn't it bloom already for you in June? The reason for this seemingly endless run of flowers is what we call *succession planting*. Instead of planting all your seedlings at once just after your last frost date, you plant out numerous times during the growing season. Here in Maine, I have three successions, which means I plant my seedlings in mid-May, in mid-June, and in late June. This means that once the mid-May snapdragons have bloomed and been harvested, the second succession is just coming into its own and is ready to be picked a few weeks later. Growing flowers this way is not just for flower farmers with acres of land. Your successions might only have five to perhaps twenty blooms, which is exactly how I started experimenting with this style of growing.

For this to work effectively, it is paramount to create a clear schedule for starting and transplanting seedlings. Over the spring, this schedule will become your nearest and dearest friend. And by the season's end, it will be memorized. While creating this schedule can be tedious and time-consuming at first, the good news is that you really only need to do it once! Subsequent years might have little tweaks and adjustments, but the bones will remain the same.

Use Your Calendar

Consider purchasing a large desk-size calendar to organize your planting schedule. To begin, create a list of all the annual flowers you plan to grow this season. I quite enjoy making this list on a bit of watercolor paper and doodling little renderings of each bloom. This makes the process seem more carefree and enjoyable, versus task-driven. Remember to select a mix of focal, secondary, and filler flowers, plus foliage. Also, keep in mind what perennial flowers, branches, or shrubs might already be available in your growing space at certain times of the year. Gather the following information for each annual flower you plan to grow:

- » Number of seedlings to be planted
- » Location in the garden
- » Spacing guidelines
- » Seed start date
- » Approximate transplant date
- » Approximate harvest date

Next bring your attention to the calendar and record your anticipated last frost date. To be safe, move three to five days beyond this date, and circle it. This date will now be known as Succession Planting One Transplant Date. Walk your fingers along the calendar and locate the date three weeks beyond Succession Planting One Transplant Date, circle it, and write Succession Planting Two Transplant Date. Use the same three-week interval to find and circle the Succession Planting Three Transplant Date. If you have a long growing season, you may be able to fit in more than three successions.

Next, record the date you need to sow the seeds of each flower variety by looking at the back of your seed packets. This will inform you how many weeks *before* your transplant date you need to start the seeds indoors. Most seed packets give you a two-week window to start seeds, and four to six weeks or ten to twelve weeks before the last frost are common intervals. After lots of trial and error, I would encourage you to be patient and start your seedlings at the later suggested date. For the examples above, this means sowing your seeds four weeks or ten weeks before the last frost. When I started seeds earlier, they often became tall, leggy, and spindly. When transplanted, these blooms are not strong or happy when they arrive in the soil.

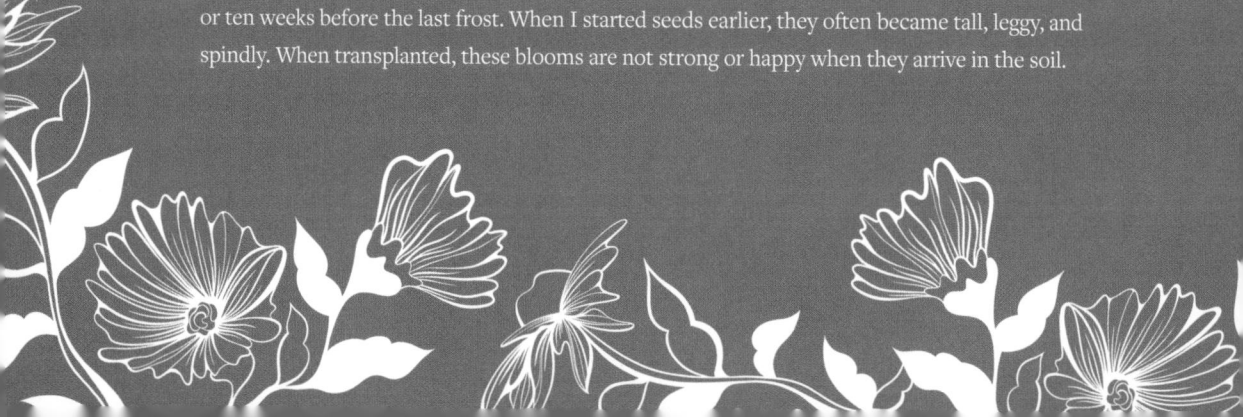

Closely read the seed packet and determine if any of your blooms require special treatment to germinate. Some of these recommendations include:

» **Cold required to germinate:** Certain flowers require a bit of cold to wake up and begin to grow. A great example of this is larkspur. If you are not directly sowing these seeds in cold soil, the best thing to do is place the seeds in the refrigerator for one to two weeks prior to sowing. Also, these seedlings should not be placed on a heat mat.

» **No light to germinate:** Some flowers need a period without light to germinate. These seeds should be kept in a dark closet until germination occurs.

» **Needs light to germinate:** Some flower seeds are so very small that if you tuck them under the soil, they will not receive enough light to germinate. Snapdragons, foxgloves, ageratum, and poppies are examples. It is best to lay these seeds on top of the soil and dust them with a small amount of vermiculite or sand. This is called surface sowing.

» **Does not transplant well:** There are flowers and ornamental grasses that do not appreciate being started indoors. For these blooms, plant them out when the soil has warmed.

I also like to record the possible harvest date for each bloom on my calendar. This can be discovered by once again returning to the back of the seed packet and looking for the metric often labeled "seed to bloom." This is the number of days or weeks it takes each bloom to grow from a tiny sown seed into a blooming flower.

Final notes: Don't be too rigid! Your successions will change as you grow seeds. Consider sowing more seeds than you need; this way if the germination rates are particularly poor for a single variety, simply plant out another flower that was a great success. Anticipate failures and be nimble and lighthearted when they arrive. This is easier said than done but it's definitely worth repeating! You are not perfect, no one is, so don't hold your garden to that standard of perfection. Messiness is part of the process, and it's fine.

Starting Your Own Seeds: The Basics

I'm going to be honest: seed starting is where dreamers often find their inspiration begin to wane. We move from beautiful floral images and fantasies of gorgeous gardens into the harsh reality of power strips and timers. The romance of the garden has seemingly vanished into thin air. But the flowers will not bloom without a proper start! And while I understand seed starting is not for everyone, I encourage you to give it a go, if not your first year, then maybe the second. Not only do you save money in the long run, but having something green to care for in the early days of spring is incredibly rewarding. Nothing is better than holding up a long-stemmed snapdragon midsummer and remembering how very tiny the seed felt between your fingertips only months before.

When I first started growing from seed, I had one small grow light, a heating pad, and a boot tray. I would start my seedlings on the heat mat on top of the dryer. Once the seeds germinated, I transferred the seedlings to the boot tray, and turned on the grow light. A bare-bones but highly effective seed-starting method.

If this is your first year growing from seed, start small! Consider a single grow light and heating pad. Using the soil-blocking method, which we will review shortly, you can easily start two hundred flowers on a small bookshelf.

Below is a list of must-have seed-starting supplies.

- Soil-blocking kit
- Vermont Compost Company Fort Vee or other organic potting soil mix
- Grow lights
- Germination heat pads
- Five-gallon bucket
- Spray bottle
- Shelf
- Timer
- Plastic or old shower curtain
- Plant tags/markers
- Pencil
- Toothpicks

SETUP

Always, always place a shower curtain or piece of plastic below your growing area. It can be difficult to clean up water and soil beneath a shelf of tiny seedlings.

POWER

I use a large surge protector and a timer. Most seedlings need sixteen hours of light every day. Even with the best intentions it is possible to forget to turn on or off the lights. So put the entire system on a timer and reduce your efforts.

WATER

A traditional small watering can will do. I give my seedlings a soak each morning. In the evenings, if any seedlings have dried out, I use a spray bottle to dampen them.

LABEL

Don't forget to label each seedling with the name of the flower and when it was sown. Label with a Sharpie or permanent marker as your workspace is often messy when starting seeds! It is very easy to confuse flowers when you are a beginner gardener. Within a few years, the seedlings will look markedly different, and it will be hard to recall how you ever confused a zinnia with a cosmos. Until that time, label, label, label!

Soil Blocking

Over the past few years, I have committed to a style of seed starting called soil blocking. Eliot Coleman, a beloved organic farmer and educator, brought this method from Europe to the United States in the 1970s. Simply, seeds are grown in small cubes of soil, not plastic trays. These cubes are created using a special soil mixture and a small handheld tool called a soil blocker. Soil blockers come in various sizes, but I mostly use the twenty-block version. This technique may seem complex for a beginner, but I actually find it is the simplest and most effective method of starting seeds, and I highly recommend it for someone who has yet to sow a single seed.

Other benefits of soil blocking include:

» **More flowers in less space:** The smallest soil blocker can create twenty flower seedlings in tiny ¾-inch squares. With a single grow light and heating pad, you can easily start an entire garden on the top of a table or dresser.
» **More affordable:** Soil blocks require less soil and thus are more affordable than larger traditional seed-starting systems.
» **Easier to water:** I find it is more obvious if you have over- or underwatered seedlings when you use soil blocks.
» **Less waste:** Forget lots of plastic trays and pots. Soil blocking is self-contained.
» **Strong transplants:** One of the major benefits of soil blocking is the so-called air pruning that occurs. Essentially, when a root hits the edge of the seed block and is exposed to the air, it begins to grow additional roots inside the soil block. When you transplant your soil blocks outdoors after the last frost, the roots are not root-bound against a pot, and thus establish more easily.

Creating Soil Blocks

To begin, take a five-gallon bucket and add water until the bucket is roughly one-quarter full. Add a soil-blocking mix to the bucket. Some people will make their own soil-blocking mix, but I've always found purchasing soil that specifies it is for soil blocks keeps things easy. These mixes are designed to hold the soil tight together, as the blocks are freestanding and need to function without the support of a plastic seed tray. Mix the soil and water well with your hands. The soil is the perfect consistency when you can squeeze a small amount of dirt in your hand, and water drips between your fingers.

Next, put three to four cups of this mixture in a paint tray, grab your soil blocker, and slowly wiggle it back and forth, filling each block with soil. Flip the soil blocker upside down and pat the soil into place. Press the soil block onto a tray and pull it up slowly. A simple cafeteria tray will hold roughly six blocks from a twenty-cell soil blocker. Fill the tray, and then begin to sow each seed, either on top of the soil or pressed in with a toothpick or pencil. Cover the seedlings with burlap or a reusable humidity dome.

Place the blocks on a heat mat (unless they require cold to germinate) with no light source. When the seedlings emerge from the soil, remove them from the heat mat and place them under a grow light. Be sure to keep your seedlings close to the grow light, roughly one to two inches. As your seedlings grow, raise your grow light. Use a timer to ensure your flowers receive sixteen hours of light each day.

Water seedlings daily by pouring water directly into the sides of the tray. Over the first few weeks, your seedlings may dry out quickly. Check your soil blocks in the early evening and use a spray bottle to dampen the soil as needed.

Designing Your Cutting Garden

With a clear idea of what flowers to grow and when to plant them, we are now ready to design your cutting garden. If this is your first time designing a garden bed, let me assure you that there is no easier place to start than with cut flowers. *Design* may sound like an intimidating word, but it is really very simple, not to worry.

A cut flower garden is traditionally designed in rows, with the same variety of flower planted right next to each other. This is a great place to start as it's easier to stay organized. Remember, most of the flowers you are growing will be cut down, so don't get too hung up on the visual effect in the garden! It will be beautiful no matter what, even if you're not so into the idea of "linearity." Rows of flowers are pleasing to the soul, and the inherent shapes of each variety will give each row a huge amount of individual personality.

To determine where flowers should be planted in a garden, you first need to consider the position of the sun and the height of each flower. Ideally, your rows will run north to south. However, I have a few that run east to west. Either way, you want your tallest plants tucked in the back row, where they will not cast a shadow on the shorter flowers.

How do you determine if you should grow in rows or a container? This is all a question of time. I suggest you start where you are and think small. Does ten minutes of effort a day sound realistic? If so, then a container or pot is perfect. Perhaps you are ready to commit to a half hour? In that case, a collection of raised beds will meet your needs. If you want to work the garden from sunup to sundown, by all means, plant rows of blooms. The amount of time you make for a garden in your life is so very personal. Understanding, clearly and without judgment, how many minutes or hours you are willing to commit will help shape the scale and design of your project. As a general recommendation, don't plan a cutting garden wider than three feet. Anything larger makes it difficult to harvest blooms effectively.

When I first began designing my own garden, I had the urge to create big, tall rows of flowers. I attributed this to my desire to connect to the carefree-roaming nature of childhood, when all was big and I felt so very small (and even the smallest rosebush seemed a mountain). It is possible I was benefitting from what Jay Appleton calls the prospect-refuge theory—simply, our evolutionary pull to be tucked away, safe, with a good view of the world around us. I'd encourage you to create a little hideaway in your garden, even something as simple as a meditation bench or a cushion in between two raised garden beds. A grander approach might be planting a fast-growing shrub such as a hydrangea that you could sit underneath. Let the cutting garden give you a sense of safety and hold you during the growing season.

As you review your growing space outline and planting list, you might find yourself wondering, do I have too many flowers on my list, or not enough? Will I have space for another succession? I promise you will find a way to squeeze all your seedlings into the earth, and if they are too close, in many cases the stems will grow taller. Interestingly, some people plant sunflower seeds closer than recommended, which results in a smaller bloom.

Interplanting (or Companion Planting)

One method for growing lots of flowers or vegetables in a small space is called interplanting. Perhaps the most well-known example is that practiced by the Iroquois and Cherokee, called the three sisters. The eldest sister, corn, grows tall and acts as a support for the second sister, beans, which adds nitrogen to the soil, and the youngest sister, squash, uses its leaves to cool the soil, keep weeds at bay, and attract pollinators with its blooms. Some call this companion planting—growing certain items next to each other in such a way that results in healthier crops.

However, interplanting doesn't always have to provide a known benefit, like bringing nitrogen back to the soil or repelling insects. A reason to interplant could be as simple as a plant germinates quickly, so I will plant it next to a slow-growing bloom that is dominating a small garden bed. I encourage you to try some of the pairs listed here or create your own! A garden is the perfect place for scientific experimentation.

DAHLIAS AND SUNFLOWERS OR ZINNIAS

Dahlias are a wonderful flower but so very slow to bloom and require lots of space. While you wait for the dahlias to become realized, interplant a quick-to-grow flower like sunflowers or zinnias. By the time your dahlias are producing beautiful flowers, you will have harvested lots of stems from an otherwise empty bit of earth. A small note: do not plant the sunflowers or zinnias until a bit of greenery from the dahlia emerges from the earth. Dahlias are prone to rot, so you should not water them until they have burst through the soil.

LISIANTHUS AND STOCK OR SNAPDRAGONS

Another slow-to-grow flower, lisianthus, is planted while the ground is still cold and will slowly but surely make its way to bloom. Experiment with planting another tall line flower, such as stock or snapdragon, which can easily squeeze in next to the lisianthus.

ANY FLOWER AND RADISHES

Radishes can be sown in between rows of flowers to fill space that would otherwise be infiltrated by weeds. Radishes are quick to grow and low-lying, so they will not compete with your bloom for the sun.

SNAPDRAGONS AND CALENDULA

Calendula is a relatively low-growing flower and can be tucked into rows of snapdragons. Once your snapdragons have been harvested, beautiful calendula will be revealed and likely in bloom.

How to Create a Garden Rhythm

While your seeds are slowly growing indoors, let's focus on creating a rhythm for your garden in high summer. Some might call this a schedule, but I feel that is too harsh a word for something as blissful as gardening. Of course, the garden of your mind's eye is bursting forth with blooms of all shapes and sizes. I doubt you are imagining aphids or quack grass. Keeping up with insects and weeds is manageable if you keep to a rhythm that involves weeding once or twice a week and putting eyes on your flowers each day. The following is how I divide each day of the week in the garden during the growing season. Feel free to follow this exactly or modify it to meet your needs.

SUNDAY

Walk the garden during the evening golden hour. Touch flowers and examine stems and petals for changes. Note stages of flowers and concerns regarding plant health. Think of the week ahead, and record possible harvests and uses (e.g., a bouquet for Blythe's visit, basil for Wednesday pesto).

MONDAY

Wake as early as your body allows—sunrise is best. Prepare a bucket of cool water and grab your best snips. Harvest what is available, place blooms in the bucket, and bring them inside to a cool, dark room. Let your flowers rest all day and arrange in the evening.

TUESDAY

Today is for tidying up the beds and weeding. Catching a weed early, when it first pops through the soil, saves you lots of time later in the season. Pull by hand or use a hoe (more on garden tools in spring).

WEDNESDAY

Midweek is the time to focus on sowing new seeds. In high summer, the excitement of early spring is long gone. With actual flowers in your hand, it is easy to forget the importance of planting even more blooms. Walk the garden and see if there are any empty spots to add sunflower or zinnia seeds; these varieties both germinate quickly.

THURSDAY

Walk the garden and look for plants that could use fertilizer. I only use organic materials in the garden. Later, we will discuss soil amendments such as worm-casting compost tea and molasses. There are also plenty of premade organic fertilizers if you don't have time to make your own. You will not fertilize every flower each week, but it is wise to give cut-and-come-again blooms a little boost after harvesting.

FRIDAY

Early morning it is time to harvest again. In spring, you may only get one harvest a week, but in high summer you can count on two days of blooms. As you did earlier in the week, wake as early as possible, harvest flowers, and place them in cool water.

SATURDAY

When the weekend arrives, I like to wake up early before the heat sets in and do some additional weeding. Saturday is also a time to rest and enjoy the fruits of your labor. One can keep very busy simply puttering around the garden but don't forget to settle into stillness and watch how the natural world moves through your space.

Winter Closing Activity: Breathe

Let us return to our dreaming mind. Look at your flower list and begin to imagine how each bloom might appear in the palm of your hand. Sketch or write out a floral recipe, which is a collection of flowers you'd like to see complement each other in a bouquet or arrangement. Review your final grow list to ensure you have focal, secondary, and filler flowers. Recall images of bouquets and arrangements that inspired you. Feel free to print them out and hang them on a wall or next to your seedling shelf. Take a few moments of rest before we dig into the dirt.

Spring

Awake & Prepare

> Ever since I could remember anything, flowers have been like dear friends to me, comforters, inspirers, powers to uplift and to cheer.
>
> —Celia Thaxter

We are now transitioning from a garden of intention to a garden realized. Spring slowly unfurls in Maine. It isn't a loud pop or a sudden bloom of azaleas while you sleep, but a slow process that awakens onto itself like a patient child peeling back the rind of an orange. One morning the front door is wide open, the sweet fragrance of hyacinth gently working its way through the house, traveling to the forgotten dark corners of the closet beneath the stairs and removing the layers of gloom from your tired winter coat. By afternoon, the front door closes shut with a startling bang, as a howling rain and wind tear through the garden, knocking over daffodils and bending tulips to touch the earth.

I must confess that I used to have a rather ho-hum attitude toward early spring. As a child, it was not uncommon to complete an Easter egg hunt in muck boots between snowbanks and mud piles. The spring of the north is not all daffodils and forsythia, as one might imagine. However, my view of early spring has changed, thanks to my children, who live completely in the present. When they are told it is the first day of spring, then certainly it is. Even if spring requires snow pants, it is my children who spy the first snowdrop, my children who hear the blue jay, and my children who demand we set up dinner outside when it is 45 degrees.

You don't need to be in the presence of a child to accept that it is spring; you only need to embrace the practice of a beginner's mind. This term, popularized in the 1970s by Shunryū Suzuki's book *Zen Mind, Beginner's Mind*, essentially means being open to all possibilities and letting go of preconceived notions. If you tell yourself March is dreary and mud is no fun, then certainly that is the reality you will experience. Why not open your mind to the possibility of beauty found in the very early days of spring?

Like many of you, my beginner's mind was activated often during the pandemic. That first March, alternating between stretching our backs in the yard and hunching over a table full of remote schoolwork, the decision was made to tap the maple tree. This was something we had always planned to do, but as the chaos and pace of general life often swallowed us, the sap buckets sat collecting dust in the garage for years. But now, tethered to home and searching for just about anything to pass

the time, we huddled around the maple as my husband drilled a small hole into the bark. Over the next few weeks, a routine developed—the sensation of earth thawing beneath my feet as I trudged out early, the sound of my hot coffee spilling on the snow, the bright winter sun reflecting off the bucket, brimming and overflowing with sap. Through this ritual, I became acutely aware of the change in temperature from late March to the beginning of April—slow changes I would have never been made aware of had I not left the house and taken to the yard.

This season I encourage you to wander, no matter the temperature! Will yourself into thinking it's spring, even when the snow falls on your tulips (they can handle it). Go outside and search for signs of rebirth in your growing space. Perhaps consider downloading some garden-inspired tech on your phone. There are a variety of apps available that can help you identify plants and flowers you might stumble

upon in your own yard or walking down the street. In the early days of maintaining my garden, what was a weed or what was a flower eluded me completely. I found this technology useful in identifying each seedling, even in the earliest stages of germination.

While wandering the yard or neighborhood, look for anything that might be repurposed. March can be wild and windy—in like a lion, out like a lamb, as they say. Look for fallen branches that could function as a trellis for climbers like sweet peas or as a support stake for snapdragons and dahlias. Working with what you have makes a garden uniquely yours and reflective of place.

If you happen to have some already-established garden beds, tidy these spaces and make way for new flowers. Pick up any sticks or branches that have blown into the garden over the winter and trim dead blooms that might be in the way of your new plantings. Consider leaving the stems and roots of these blooms in place, as they can act as organic matter and help enhance your soil.

Now that it is time to get into the dirt, let's review some basic gardening tools that are helpful to have on hand.

Basic Garden Supplies

You do not need to purchase all of these items at once and if you are planning a small garden, items like a broad fork or wheelbarrow will be unnecessary. But every gardener should invest in a good pair of snips and a trowel (maybe even one with a fancy wooden handle, if you want to treat yourself).

> So deeply is the gardener's instinct implanted in my soul, I really love the tools with which I work—the iron fork, the spade, the hoe, the rake, the trowel, and the watering-pot are pleasant objects in my eye.
>
> —**Celia Thaxter**

» **Broad fork:** This is my new favorite tool; I use it in early spring to aerate the garden rows. You simply hop on top of the middle bar, wiggle back and forth on the handles, and pull to expose the new soil. It is also useful when replanting a bed midseason, for example, after I pull tulips and before I plant dahlias.

» **Buckets:** A collection of five-gallon buckets is a must in your garden. Use them to harvest flowers, transfer small amounts of soil, or hold a vase at eye level on a tabletop when you play with flowers. Inexpensive alternatives are frosting buckets from your local grocery store. Just ask the bakery to give you a call when they have any left over. In high summer, thank your new friends by gifting them a bouquet of flowers.

» **Cultivator:** Perfect for small gardens and raised beds, the handheld three-prong metal cultivator loosens soil before you plant seeds.

» **Garden gloves:** I prefer to get my hands into the dirt, but if I'm trimming roses or shoveling soil for long periods of time, garden gloves are a must.

» **Garden hoe:** Use this tool to pull up weeds as they begin to surface in your garden. It is especially useful if you have a tender back, as you can stand straight while gently working the soil.

» **Hori hori knife:** This Japanese gardening tool is a serrated knife used to remove weeds, divide plants, and cut sod. Wonderful in gardens of all sizes.

» **Japanese hand hoe:** Just as it sounds, a handheld version of the garden hoe that is perfect for small spaces and raised beds.

» **Loppers:** A long-handled pruner that can cut through thick branches your regular shears cannot.

» **Plant support:** Many flowers will require some staking and twine as they grow. Bamboo stakes are aesthetically beautiful, and jute twine can be used between the stakes to corral blooms or to wrap around a single plant and secure it.

» **Rake:** A rake in the garden has many uses, including pulling back layers of spent plants in early spring, roughing up or evening out soil, and, of course, moving things like leaves. If you plan to create a stone path or space around your garden, purchase an additional metal rake. Also, consider an eight-inch rake, which is useful in small spaces or tight corners.

» **Shears:** For pruning and harvesting flowers, one can never have enough pairs. I also keep a pair in my purse just in case some beautiful roadside weed catches my eye.

» **Spade:** The use of a garden spade is endless. It can create clear lines for edging your garden bed, dig a space for a new planting, or free up large roots.

» **Trowel:** Perhaps the most quintessential gardening tool, the trowel functions as a handheld digger, and you'll need it every day.

» **Wheelbarrow or gardening cart:** Useful in a larger gardening space to haul dirt and debris or carry buckets full of blooms while harvesting.

If modifying for injury is a concern, the following is a list of tools that encourage healthy garden habits.

- » **Ergonomic garden tools:** Designed with a curved handle, these tools reduce strain on the wrist.
- » **Five-gallon bucket:** Flipping a five-gallon bucket upside down to sit on is easy on the knees and keeps your back in a better-aligned position for harvesting.
- » **Garden kneeler:** This is a great option to protect your knees as you weed or harvest. Consider a simple pad or a kneeler with handles that allows you to slowly lift down and up from a kneeling position.
- » **Long-handled wire weeder:** Designed by Eliot Coleman, this weeding tool can be used to remove weeds from rows without the need to bend over and pull. You will have to keep up and use the tool regularly, as it doesn't often remove the entire root from your garden bed.
- » **Ratchet pruning shears:** These shears require less hand strength to cut through stems and small branches.

Developing a Watering Plan

The general rule of thumb for a well-watered garden is roughly one inch of rain per week. However, some plants are thirstier than others, and you will notice areas of your garden that require more or less water than others, depending on soil type and sun exposure. The most reliable way to determine if your garden needs water is a bit old-fashioned. Simply dig your finger into the soil about one inch and take notice of

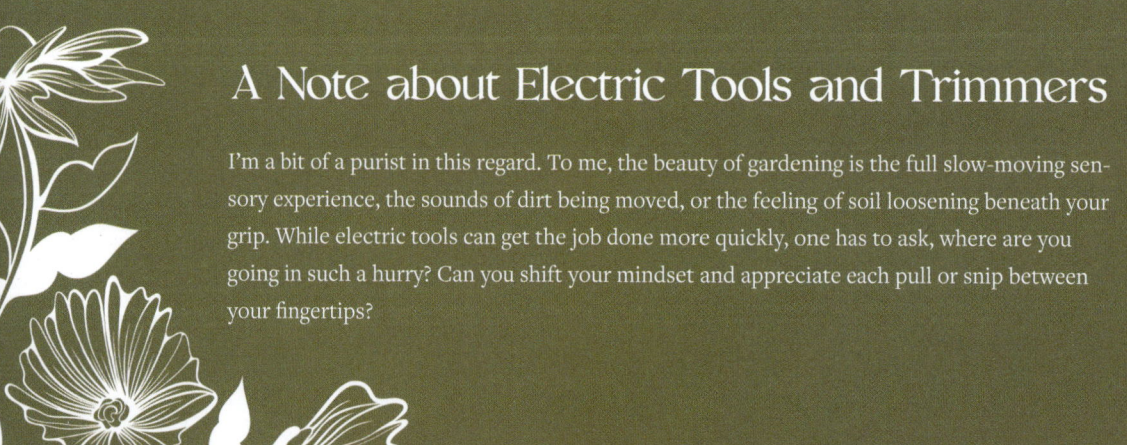

A Note about Electric Tools and Trimmers

I'm a bit of a purist in this regard. To me, the beauty of gardening is the full slow-moving sensory experience, the sounds of dirt being moved, or the feeling of soil loosening beneath your grip. While electric tools can get the job done more quickly, one has to ask, where are you going in such a hurry? Can you shift your mindset and appreciate each pull or snip between your fingertips?

the moisture content. If the soil feels damp, you don't need to water. If it is dry and crumbly, then you should water.

An important element of watering is determining what kind of soil you have. The mason jar test is a basic method to test your soil at home and determine its type. Take a large mason jar and fill it roughly one-third of the way to the top with soil from your growing space. Be sure to remove any large rocks or clumps of organic matter. Add water and a drop of dish soap, then shake the mixture. Let this sit for forty-eight to seventy-two hours; as time passes you will begin to see three distinct layers. The bottom layer will be coarse sand, the middle silt, and the top layer will be clay. Examine each layer closely, determining if your soil is well-balanced or has noticeably more or less of a certain element.

Here at the farmhouse, almost everything is clay, so the gardens need long, deep soaks, maybe once or twice a week in high summer, as the drainage is minimal. You may have to water more often for sandy soil, but not as long, as the water will evaporate more easily. A container or small raised bed will need watering each morning in high summer unless it rains.

Determining the best method for watering your garden will largely be based on how much area you are growing on. A gentle reminder: don't feel you have to make the proper decisions on a watering system today. I started with one irrigation method at the farmhouse last season, a watering can and backpack sprayer, then moved on to drip tape and completed the season with a sprinkler! To start, just select one method, and be open to the idea of pivoting midseason. There are no absolutes in the garden. Following are some ideas for watering methods.

» **Rain gauge:** Affordable and simple, a rain gauge is a single tube with measurements that help you determine how much water your garden has received in a given storm or week.

» **Watering can:** A three-gallon watering can with a long spout is just right for most gardens. I thoroughly enjoy hand-watering my garden, as it gives me time to slow down and admire each bloom up close.

» **Water wand:** Perhaps the most common method for watering your garden is a wand that attaches to your hose. Consider one with a telescoping handle and multiple watering options.

» **Battery backpack sprayer:** This is a fantastic way to water a large garden. With various hose attachments, you can use this sprayer to water your plants, control aphids (by spraying them with water), and apply thick soil amendments. It is fairly easy to use and has a rechargeable battery. A small note: I place my backpack sprayer on a picnic table and then fill it with water. This way, I don't have to pull the heavy sprayer up off the ground and strap it to my back.

» **Tripod sprinkler:** I purchased a tripod sprinkler at the end of my growing season and absolutely loved it. It ran on a timer and was movable, which allowed me to easily pull it back toward the house if we had guests or needed to mow the lawn.

» **Drip irrigation system:** This is a ground-based soaker irrigation system that stays in place all growing season. Don't be intimidated;

when I first ordered this setup, it stayed boxed in my garage for months! However, when I finally sat down and read the directions, the job was complete in a few hours. A major consideration with drip irrigation is the need for a spigot near the garden. If you don't have a close spigot, you will need to bury or stake a large mainline to run water to your site.

» **Spigot splitter:** This allows for different watering zones in your garden. It often has two or four outputs, but keep in mind that water flow decreases when you use a spigot splitter.

» **Bluetooth timer:** This is useful for creating a watering schedule that turns your drip system or sprinkler on and off. Some Bluetooth timers come with a spigot splitter, and even fancier varieties will know when it has rained and will skip a day on your preselected watering schedule if necessary. While it is tempting to employ the "set it and forget it" mantra when using a watering timer, remember that each week is different. There will be rainstorms, high heat, and wind; all of these elements alter how much water is needed day to day. Don't forget to observe your plants; if they are wilted and you haven't watered in a while, chances are it is time to do so.

The Beauty of Starting Small

In winter, I suggested you determine the scale of your garden project by considering how many minutes or hours you intend to dedicate each day to growing flowers. With this information, you likely decided whether you'd be growing flowers in a container, raised bed, or row-style cut flower garden.

The first garden I established from scratch was a single three-by-six-foot raised bed next to our kitchen door. We packed it with lettuce, zucchini, and tomatoes, doting on each plant that bloomed over the growing season. As the children ran in and out of the house, they would stuff their pockets with grape tomatoes or eat lettuce right from the dirt. The seemingly easy success of this tiny space gave me the confidence to add a few additional beds the following season and many rows of flowers within a few years.

Had I not started my garden small and failed, I'd have convinced myself I could never be a gardener. So please remember there is beauty in starting small and meeting yourself where you are. Learning the individual needs of each bloom you plant is important for success. Tending to only a handful of flowers allows one to fully appreciate the slow, almost unnoticeable changes that occur in a garden each day. Often the highlight of my morning is watching a single flower react to the sun by pulling individual flower petals slowly open toward the sky. With a large garden, these small moments of beauty can be missed.

Preparing Your Garden

It does not matter if you are growing in a single pot or plotting a large row of flowers, everyone must first begin by preparing their growing space. Taking time to thoughtfully consider the soil one uses or the way sunlight moves through a garden is an important first step in every growing season.

If You're Gardening in Containers or Pots

When it comes to pots, I love the simplicity of terra-cotta, and a collection of terra-cotta pots in various shapes and sizes tucked in a sunny corner of a patio is lovely. However, there are endless options of vibrantly colored and patterned pots that can brighten a growing space. Just make sure you have ample drainage holes in the bottom of your pots and that they are filled with the proper soil. We'll talk more about this later.

I fill my larger garden pots with an array of flowers in different textures and colors, similar to how I craft a bouquet. Take into consideration whether the pot or container will be viewed from all angles or just one side. Be mindful of plant height and consider flowers or herbs that gently cascade over the edges, such as creeping thyme or nasturtium. Focus on cut-and-come-again flowers such as zinnias or cosmos to keep interest all season. These varieties will churn out lots of stems for your table and constantly appear lush and in bloom through the growing season.

If You're Using a Raised Garden Bed

I would suggest you choose garden bed material that matches the general aesthetic of your home. Galvanized steel and cedar are my favorites, but spare wood or branches held in place by a bit of rebar can be an affordable alternative. To create a raised bed on top of your lawn, line the bottom of the bed with cardboard or landscape fabric to prevent weeds from growing between your flowers. In an effort to reduce the amount of soil you need, consider adding rocks, sticks, leaves, or other organic material at the bottom of your garden bed. One only needs about eight to twelve inches of soil to grow most annual flowers. In my very first raised bed garden, I spent a fortune ensuring the entire bed was filled top to bottom with soil! This was completely unnecessary.

If You're Using an Elevated Garden Bed

An elevated garden bed looks like a small table filled with soil. These are fantastic for those of us with lower-back pain or stability issues. The bed tends to be waist-high, thus reducing the need to bend down to the earth over and over again. Simply stand or sit in a chair alongside the bed and tend to your garden. Prepare an elevated garden bed the same way you do a simple raised bed, but fill it entirely with high-quality soil, as most of these beds are quite shallow.

If You're Practicing No-Till or No-Dig

Popularized by the British gardener Charles Dowding, this method of growing involves developing a garden right on top of your grass without disturbing or digging into the soil. The row garden where I cultivate the majority of my flowers has been prepared using the no-dig method.

The Basics of No-Dig Gardening

First, measure and stake the desired growing space and trim or mow the grass. Then, lay down cardboard, preferably without ink, over the area you have staked out. Spray the cardboard with water and cover it with approximately three inches of compost. Keep this area continuously wet for two to three weeks, allowing the cardboard to break down. After a few weeks, return to the site and add leaves, sticks, or any other mulch to build up the height of the garden. Oftentimes I'll top off the new garden row with a layer of loam. Some gardeners plant directly into the compost without adding other layers on top. If you plant directly into the compost, just ensure it is not still active, as this can kill your precious seedlings. Compost that is active, also called "hot," will often smell strongly and steam.

After the garden bed height is to your liking, it is time to plant your seedlings. You will need a good knife or even a screwdriver to poke through the cardboard layer, ensuring that the roots can access the soil below.

Determining Your Soil Needs

When it comes to cultivating a garden, the most important element lies beneath your feet: the soil. One way to carefully control your soil is by using raised beds and adding a top-quality organic soil mix. However, for those of us growing directly in the earth, it is important to take a soil test each year. This will reveal what types of amendments are necessary to create an ideal growing environment for our flowers.

So, what makes good soil? It is a mix of organic matter (compost, plant debris, leaves), minerals (clay, silt, sand), air, and water. Heavy clay soil will hold nutrients well but has poor drainage. For soil that is sandier, the opposite is true: water drains well, but the nutrients will not hold as readily. Learning in great detail the strengths and shortcomings of your soil is essential to maintaining a cut flower garden.

I suggest contacting your local state extension program and ordering a soil-test kit for a more in-depth analysis. These are often affordable and provide valuable information about the health of your soil. If you have a few different garden locations around your yard, take samples from each site and be sure to label them!

TAKING A SOIL SAMPLE

To take a soil sample, dig six inches into the ground for accurate results and collect your sample. The top layer of soil often holds more nutrients, so digging beyond this point will give you a clearer analysis. Remove any rocks, leaves, or sticks that end up in the sample. When the sample returns, you might be overwhelmed by the data and recommendations, so let's break down some of the numbers and terms you might see. What your soil sample tells you:

» **Soil pH levels:** This will be listed as a whole number. The ideal soil pH level can differ depending on what you grow in a given bed. For example, blueberries and azaleas like more acidic soil, with a pH of 4.5 to 6.0, but for the vast majority of plants, you'd want your pH to be between 6.0 and 7.0.
» **Percentage of organic matter:** This number should be between 3 and 6 percent and represents the amount of living and dead things in your soil. Not enough organic material means your flowers won't have the correct nourishment, and too much can actually harm your plants.
» **Breakdown of the major nutrients in organic matter:** These include phosphorus, potassium, calcium, and magnesium. The soil test will likely describe if you have a low, medium, optimum, or above-optimum amount of each nutrient.
» **Micronutrients:** Your soil will be evaluated for boron, copper, iron, manganese, and zinc. Boron is an important micronutrient for cut flowers as it impacts root and bud growth.

AFTER THE SOIL TEST, NOW WHAT?

After reading all the data about your soil, deciding what to do next can be confusing. Most soil samples provide soil-amendment recommendations, complete with the desired amount and time of year to apply. Autumn tends to be the best time to apply soil amendments like lime, compost, wood ash, and bonemeal. Having a soil test completed in the fall and then spreading the recommended amendments before the ground freezes will speed up the planting of your seedlings come spring. Therefore, put a reminder on your calendar each September to take a sample. It is satisfying to see the positive changes you have made to the garden's soil health year after year.

An Exercise in Grounding Your Body

Before adding your soil, focus on how your body moves. Remember to bend your knees and sink into your feet, grounding into the earth below you. Breathe in as you push your shovel into the pile and notice the strength of your abdomen as it tightens to lift the soil up toward your body. Exhale, forcefully—and with sound if you are brave—as you place the dirt into a wheelbarrow or walk it toward your new garden.

Looking at the entirety of a mound of compost or soil can be intimidating. Focus on the simple, small task of digging, the smell of the soil, feeling the weight of the shovel in your hand, and the strength of your lower body. Instead of dreading the task of moving dirt, can you shift your mindset toward gratitude?

Getting Your Soil and Compost

One method for obtaining soil and compost for containers or smaller raised beds is to purchase one- or two-cubic-foot bags from a favorite local greenhouse. There are a variety of soil mixes to choose from, but organic soil and those marked for raised beds or containers are best. These soil mixes are designed to hold water and keep air around the roots of the plant. Do the math to find how many bags of soil in cubic feet you will need.

For a small raised bed, figure:

length x width x height (in feet) of your bed = total cubic feet of soil needed

If you plan to maintain a few beds or a larger garden, consider having your soil delivered by a reputable landscaping company. You will be surprised how quickly the bulk option becomes affordable. When ordering, you will need to decide whether to order loam, compost, or a mix of both. Loam is soil created to be the perfect blend of silt, sand, and clay. Compost is recycled organic material added to enhance your soil. Compost from a landscaper is often sourced from yard waste, while other compost companies take food waste from local restaurants and businesses and, a few months later, are able to deliver this material directly to your garden.

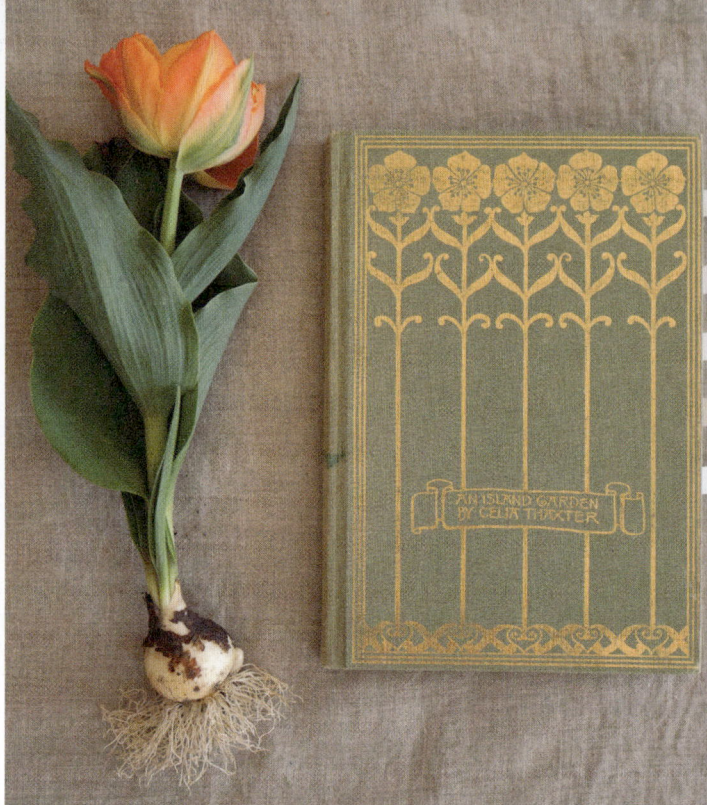

When you create a new garden, it is important to primarily use loam, as a garden full of only compost will be too rich and will kill tender seedlings. I mostly use compost to freshen up the gardens each season by applying an inch or two to already-established garden beds that look tired.

To complicate things, when soil or compost is delivered, it does not come in cubic feet like smaller bags of soil, but often in cubic yards. To determine how many cubic yards of soil or compost you need to order, use this equation:

length x width x depth (in feet) of growing space divided by 27

The Beginning of Flowering

Each spring, I record in my gardening notebook when the very first snowdrop, spring's first flower, arrives. Most often, it bursts through the earth bringing with it a flood of excitement for the garden that awaits. The first flowers of spring are always bulbs, and they could not be easier to cultivate. Simply plant them in the soil in late fall and watch them bloom to signal the awakening of your garden.

Identifying Your Favorite Spring Bulbs

Perhaps the most famous bulbs are tulips and daffodils. But might I ask that you consider a few more varieties to add to your pot or the soil? *Galanthus*, or snowdrop, tends to push through the melting snow just as the birdsong returns. Their delicate white petals dip toward the earth, almost bowing to the frozen soil below. Harvesting them feels nearly criminal, so I only cut one and keep it in the most decadent bud vase I can find in a high-traffic area where it gets plenty of attention. I place mine next to the kitchen sink, where one can feel stuck in never-ending cycles of dishes and to-do lists. Staring at that tiny little bloom reminds me of the incredible beauty in the rhythm of seasons and life. If you change the water each day, a snowdrop will last roughly a week—impressive vase life for such a tiny flower.

Grape hyacinth (*Muscari* spp.) is just slightly larger than the snowdrop. Its miniature purple blossoms always remind me of Hans Christian Andersen's children's tale *Thumbelina*. It is as if the heroine, a tiny girl who befriends fish and butterflies, could poke her head out between a collection of these blooms in the soil at any moment. Its fragrance is truly intoxicating, especially after a spring rain.

On that note, hyacinth (*Hyacinthus* spp.) is perhaps the most fragrant of all spring bulbs. There is no mistaking when this flower is in full

bloom, as the sweet smell rolls through open windows with ease. The tiny blossoms, in shades of pink, blue, or purple, are almost bell-shaped and completely hide the long, tall stem.

The genus *Fritillaria* is not known for its smell but for its artful shapes and colors. There are many species of this flower. Some, like *F. persica*, are tall, thin line flowers, while *F. meleagris* (also known as checkered lily) has small bell-shaped flowers with petals in a deep purple check pattern that are simply gorgeous.

A Bounty of Daffodils

There are thousands of daffodil (*Narcissus* spp.) varieties and, in fact, I have found myself overwhelmed by the seemingly endless options of color and shape. A daffodil's given name often reflects its size, shape of petals, and the cup in the center of the bloom known as the corona. Daffodils are an absolute obsession for some, and should you find yourself leaning in that direction, the American Daffodil Society is a wonderful resource.

HOW TO GROW DAFFODILS

First, plan ahead. The most popular varieties of daffodils are often sold out when you begin to think of planting them in the fall. Take some time to look through a

My Favorite Daffodil Varieties

'Apricot Whirl': A single vibrant row of exterior white petals, with a ruffled peach-toned split-cupped corona. This bloom made me fall in love with daffodils.

'Ice Follies': A long-lasting daffodil after bloom. It has yellow ruffled center petals and delicate bright white exterior petals.

'Professor Einstein': Giant blooms, with a large orange center and bright white exterior petals in a gentle starlike formation. The colors always remind me of a fried egg!

'Sir Winston Churchill': Delicate little white petals with a vibrant yellow center, four flowers per stem.

bulb catalog, decide which varieties speak to you, and order them in spring. Personally, I enjoy the cream- and peach-colored daffodils best.

Next, plant your daffodil bulbs two to four weeks before the ground freezes in late fall. Find a sunny spot with good drainage. Space the daffodil bulbs three to four inches apart, with the pointy end facing up. Water the bulbs once after planting, and then leave them for winter.

When the daffodils emerge, I harvest roughly half to bring inside and keep the rest in the ground. You should cut flowers just before the petals burst open for the best vase life. Something to note is that daffodils are what we call a "dirty flower," which means they tend to gather bacteria and slime on the bottom of their stems and contaminate the water for the rest of the blooms. Other dirty flowers include zinnias, marigolds, yarrow, snapdragons, and dahlias. If you arrange with any of these flowers, be sure to keep cutting their stems and change the water daily.

After the daffodils die, *do not cut back the leaves in the garden*. Photosynthesis must occur through the daffodil leaves to create energy and more bulbs beneath the soil. Once the leaves have yellowed, however, you may trim and tidy the area if you'd like.

Finally, yes, daffodils come back year after year after year! Also, as daffodils are slightly toxic, groundhogs and deer will not eat them, making this bloom a wonderful investment for your garden and your spirit.

Tulips Galore

I love tulips! The real difficulty in growing them is selecting the color palette, as there are so many incredible varieties to choose from. What variety, shape, texture, and tone dance with each other? In previous years I've pulled color inspiration from a favorite painting. Most recently it's been *Poppies, Isles of Shoals*, by Childe Hassam, which depicts a vibrant field of poppies planted lovingly by Celia Thaxter. The contrast of the red, pink, and white blooms against the dark, iron-rich rocky coast is breathtaking. With a print of the painting in one hand and a cup of tea in the other, I spent a dreamy winter afternoon poring over bulb catalogs and comparing the tone of tulip petals to the blooms Hassam painted many years ago. This meant that in spring I was greeted with an array of tulips that worked together harmoniously in a vase.

My Favorite Tulip Varieties

'Angélique': Very different in color from all the other tulips I grow, this light pink bloom has plentiful petals and is a classic.

'Black Parrot': A long-stemmed, ruffled-edged single-bloom tulip. It is similar in color to another favorite variety, 'Palmyra' (see below), but a bit darker, almost black in certain light and more textured.

'Dream Touch': The deep purple petals are accented with a white edge that is so perfect it appears as if someone drew it on the flower.

'El Niño': This single late variety has incredibly long stems, with colors similar to those of the double early 'Orca'. The single petals are long and elegant, with pointed edges.

'Orca': The stem length of this double early variety is nothing to write home about, but the color is fantastic. Golden-orange petals are highlighted with green lines running up each petal's folds.

'Palmyra': I adore a deep-colored tulip, even though I find it quite difficult to photograph! As a double early, these will bloom in time for Mother's Day, even in the far north. Lots of layers of deep burgundy petals.

'Parrot King': This bloom looks stunning as it opens over the course of a week. Its petals spread to an enormous width, exposing vibrant yellow and orange tones.

'Queensday': Gorgeous deep orange petals that work as a wonderful contrast flower to pink or darker purple tulips.

Angélique

Black Parrot

Dream Touch

El Niño

Orca

Palmyra

Parrot King

Queensday

Since I grow tulips to sell, when harvesting I pull the entire bulb out of the earth. This ensures a long stem length and gives me space to plant a new crop. As suggested by Bridget Elworthy and Henrietta Courtauld in *The Land Gardeners: Cut Flowers*, I rotate dahlias and tulips in the same bed. When the tulips come out, the dahlia tubers are planted. The reverse occurs in autumn. After the first killing frost, I remove the dahlias and plant the tulip bulbs. Of course, this means your tulips function as an annual bloom, and a new crop of bulbs must be planted each autumn. If you are not planning to remove the tulips completely, so that they may return every year, the challenge is what to do with the precious garden space after the tulips bloom. Planting flowers over your tulips will likely require water, which could lead to the rot of the bulbs. If you're going to leave your tulips in the ground, I suggest planting a drought-resistant annual on top, such as yarrow, as you can avoid watering it altogether and rely primarily on natural rainwater throughout the season.

PLANTING TULIPS

To plant a handful of tulips, a bulb planter allows you to simply hollow out a bit of soil and drop a single bulb into place. If you ordered thirty bulbs or more, I suggest either the trench or the no-till tulip-bed method. In the trench method of planting, you dig a long, narrow trench in the ground at a depth of six to eight inches. Usually I dig a trench three by six feet, which allows me to plant hundreds of bulbs if I set them in tightly, near to touching. Be sure to place your tulips so that the narrow tip faces up. After planting the bulbs, cover them with soil that reaches ground level and give them a good soak of water.

The no-till tulip-bed method is similar to the trench style of planting, however, you will not dig into the soil. This is especially useful if you have a tired back after a long season of gardening! For the home gardener, I would recommend purchasing a small raised bed or constructing a raised bed from spare branches or wood held together with rebar. Put a thin layer of loam and compost at the bottom of the garden bed, place your tulips close together but not touching, and cover the bulbs with six to eight inches of soil.

Adding Shrubs

While spring is still too cold for many of our precious annual seedlings to make their way out into the soil, you may want to consider adding some shrubs or trees to your landscape, as woody stems from shrubs provide solid structure in a bouquet or arrangement, holding up heavy focal flowers. As the leaves of many of these shrubs often change color throughout the season, I find they tell the visual story of local seasonality like no other.

Shrubs to Consider

The perfect time to plant new shrubs is once the ground thaws and dries out but before things begin to bloom. As many of the annuals I grow are not natives, I have added more native shrubs surrounding the cutting garden and the yard's borders. The bees and butterflies are pleased. But the best part about shrubs and trees is that they require little care once established. I'm always looking for less effort in the garden. Here are some shrub varieties I recommend.

ARROWWOOD (*VIBURNUM* SPP.)

Native to the eastern United States, this shrub provides interest all season long. White lacelike flowers in spring make way for deep blue berries in summer, and the constantly shifting leaf colors can be used as foliage all season long. Plant in moist soil, in full sun to part shade.

HYDRANGEA SPP.

It took me a moment to come around to hydrangea, which I long associated with fancy second homes dotting Cape Cod. While I haven't planted a single classic blue hydrangea, I do have vibrant *Hydrangea paniculata* 'Limelight', a variety that glows under the moonlight and is perfect when dried. Climbing hydrangeas are lovely for maximizing vertical space, and lacecap hydrangeas grow quickly with lush greenery. Be sure to research your hydrangea variety closely, as many have different sun-exposure requirements and need to be pruned at different times of the year.

LILAC (*SYRINGA* SPP.)

I don't think there is any flower scent that transports me like that of a lilac. It represents windows open, spring air on winter skin, and the hope of summer. Make sure to plant your lilac in well-drained soil in full sun and prune immediately after it flowers; if you wait too long to prune, it will have set new buds for the following season, and you'll end up pruning off the buds and missing a year's worth of bloom time. So, prune after bloom!

LOWBUSH BLUEBERRY (*VACCINIUM ANGUSTIFOLIUM*)

Perhaps one of the most famous natives in my state of Maine, but also native to many areas in the Northeast and Canada. This shrub appreciates full sun, well-drained acidic soil, and grows to roughly two feet high. It is wonderful in bouquets for structure and foliage.

NINEBARK (*PHYSOCARPUS* SPP.)

Native to North America, the cultivated variety of this shrub produces stems dotted with purple-burgundy leaves that make vibrant focal flowers like peonies or dahlias simply pop in a bouquet. The strong woody stems also function as structural mechanics for an arrangement, and in late summer you'll discover they are covered with delicate little blooms. This shrub is hardy and likes sun and well-drained soil.

wbush Blueberry

Ninebark

NORTHERN BUSH HONEYSUCKLE (*DIERVILLA LONICERA*)

An easy-to-grow native shrub that provides a lovely bit of filler flower or foliage. You will notice small yellow blossoms in summer and red-leaved foliage in fall. Cut from this bush all season long and plant in the shade.

SMOKE TREE (*COTINUS* SPP.)

An enormous smoke tree lives on the main street in our tiny little town, and each spring it stops me in my tracks. This stunner grows well in partial shade and puts out billowing smokelike flowers in early summer. When the flowers pass, the leaves can be used as foliage for the remainder of your growing season. Plant in full sun in well-drained soil.

STAGHORN SUMAC (*RHUS TYPHINA*)

This thrives in rocky soil, which is why it was so plentiful on the dirt road to my grandmother's house by the sea. Plant in well-drained soil in the sun. I harvest the berries for natural-dye projects in summer and winter, but make sure to leave lots of them behind for the birds.

Hardening Off Your Seedlings

As early spring wanes, the days will grow noticeably longer, and the soil slowly warms. When your last frost date approaches, it will be time to plant your precious seedlings out in the wild. Gradually introducing them to the world outside before planting them directly into the soil is paramount for success. For weeks, and in some cases months, they have been lovingly cared for in a stable and controlled environment—just the right amount of light, water, and, of course, no wind. Were you to pull them straight from indoors and plant them into the earth, they would likely suffer what is known as transplant shock. A flower in transplant shock will droop, wither, and ultimately die. To prevent this from happening, slowly introduce your seedlings to the world outside your door.

First, locate a shady, covered spot in your yard. This could be under a porch, in a garage with the door open, or against the side of the house. The idea is to find a location protected from direct sunlight and wind. If you don't have such a spot, create one by draping a sheet or towel over the back of a chair. Get creative!

Bring your seedlings out into this environment for four hours a day to start. When you retrieve them, you might notice that some seedlings have thrived outdoors, while others look near death. Provide them all with a hearty drink of water and let them rest indoors under the grow lights. You will be surprised how quickly they perk up.

Observe which flowers seemed to hold their own outdoors. Consider pushing these blooms along more quickly in the hardening-off process. Of course, the opposite is true for the seedlings that struggle; they may need additional time protected in the shade. Each day, gradually add a few more hours of natural outdoor exposure to each set of seedlings.

The whole process of hardening off should take about three or four days, depending on the health of your seedlings and the plant variety. You may recall some seed packets state that a flower "does not like being transplanted." Remember to treat these blooms with the utmost care, moving them slowly along toward the soil.

By far, the most challenging part of this entire process in terms of plant vitality is the final day, in which the seedlings spend a full day in the sun (note: if your bloom requires shade, skip this step and simply harden off in the shady spot you

plan to plant them). It may be that you need to water the seedlings at midday so they don't dry out.

A few additional words of advice: Avoid hardening off in the rain! Soil blocks break down if they are watered too heavily. Overcast days are fantastic for gradually introducing your seedlings to the world. If you work outside of the home, start this process on the weekend so you are close by to monitor them. Finally, some seedlings will die, even with your best efforts. This is to be expected. Try not to get too disappointed.

A Word on Planting Dahlias

Dahlias are native to Mexico and Central America, where they are perennials, meaning they come back year after year. For those of us with deep winter, dahlias must be dug out each fall and stored in a cool, dark place over the winter, otherwise they will not survive. A dahlia is grown from a tuber, which looks like a small potato or other root vegetable. A true cut-and-come-again flower, dahlias seem to begin blooming as all the other flowers start to tire in the late-summer sun. These lovely plants will produce blooms right up until the last frost. Dahlias are deer-resistant, but they are not deer-proof. Still, they are a beautiful bloom to try for those worried about wildlife.

When the lilacs have bloomed is Mother Nature's way of saying it is time to plant dahlias. First, you'll want to inspect your tubers for signs of rot and damage. Ensure that you can locate the eye, which is a small, usually white dot at the neck of the tuber. Plant the dahlia in full sun four to six inches below the soil, horizontally, with the eye facing upward, and approximately nine to twelve inches apart. Make sure to plant them in well-drained soil, otherwise the tubers will rot. One particularly rainy spring I lost hundreds of dahlias in a low-lying section of the garden. Since that event I mostly plant them in raised beds or pots.

Do not water your dahlias until little green bits of growth have emerged from the soil; again, tubers are prone to rot. A bit of rain is nothing to worry about, but it is important that you avoid saturating the tubers with water. Dahlias grow quite tall, and therefore benefit from stakes and string to hold them in place. Put stakes into the ground earlier than you think. It is difficult to stake the blooms once they have begun to grow.

Reminder: since dahlias take up lots of space and are slow to bloom, try interplanting them with sunflowers or zinnias.

The Process of Pinching

Flowers that we call cut-and-come-again—those that will continuously produce flowers after being cut—benefit from a process called pinching. To pinch a flower, take a clean pair of snips and cut the center stem of your flower just before it blooms. Usually this means after a given flower has put on three or four sets of leaf nodules and a flower bud is visible. This can be a painful process for a beginner gardener; after waiting months for a single bloom, you are directed to remove the flower, which seems counterintuitive to the growing process. However, in doing so, you will produce many more flowers, and their stems will be longer. Pinching flowers will delay the bloom by one to three weeks, depending on the flower variety. Oftentimes I will avoid pinching a few of each flower variety, just so I can enjoy them myself sooner rather than later. This is about pleasure, is it not?

Harvesting Spring Flowers

When I began selling flowers, people often remarked, "Your garden must be so beautiful," to which I'd reply, "Well, not exactly." Most people don't realize that when you grow flowers for the purpose of cutting, the flowers are harvested before they burst open into bloom, which ensures a long vase life after being cut. In fact, when photographing a cutting garden, the rows of lush blooms are usually a bit of smoke and mirrors. Anyone growing flowers for profit would cringe at the site of a flower field full of unfurled petals and bright, vibrant colors.

However, for a home garden, you can cut a flower whenever you like. Your garden is yours, after all. If observing a flower transform from seedling to bloom to seed pod under the big blue sky makes your heart happy, then do just that. Personally, I have a soft spot for flowers that have begun to curl their leaves inward and decay, the final swan song, if you will. For many of my arrangements, I often pick and design with flowers at different bloom windows, some with buds closed up tight, and others opened wide near wilting, illustrating the life cycle of a flower.

When I am cutting flowers to sell or give away, I tend to follow the rules of harvest windows for the best vase life. Following is a list of harvest supplies and the suggested time to harvest flowers. Harvesting flowers at these stages can seem a bit intimidating at first. However, within a few days, your blooms will begin to open inside your home and your confidence will build.

My Favorite Harvest Tools

» **Floral shears (or "snips"):** I highly recommend two pairs of snips, one junky pair that you don't mind misplacing and the other a high-quality pair you treat with the utmost respect. The brighter the color of the handle, the better; I can't tell you how many times I've been snipping flowers, found myself daydreaming, wandering halfway around the yard, only to return and have no idea where I left my snips. The bright red handles have saved me over and over again.

» **Holster, hip bag, or garden tote:** Keep all your harvesting tools in a clean bag or tote you can bring around the garden. Your trowels and cultivators often become covered in dirt as the season progresses, even when you promise yourself you will wipe them down at day's end. Keeping harvest tools in a separate bag or wrapped in a soft piece of cloth ensures they stay tidy.

» **Buckets:** I try to keep the buckets I use for moving garden dirt and hauling items into the compost separate from the buckets I use to cut flowers. This limits the chances of spreading disease or bacteria to newly cut blooms. Buckets need to be cleaned by washing with soap and water before putting freshly cut blooms into them.

How to Harvest Your Flowers

Go out early, at first light, or, if you are a night owl, at dusk, to harvest your blooms. Harvesting in full sun at midday is not recommended, as it will cause your flowers to wilt. Make sure your buckets have been thoroughly cleaned and filled with cool water. If you have flowers such as daffodils or zinnias in bloom, which are known as dirty flowers as they can make water mucky, make sure you have an additional bucket and harvest them separately.

> I go forth between five and six o'clock to cut them while yet their gray-green leaves are hoary with dew, taking a tall slender pitcher or bottle of water with me into the garden, and as I cut each stem dropping the flower at once into it, so that the stem is covered nearly its whole length with water; and so on till the pitcher is full.
>
> **—Celia Thaxter**

For most flowers, snip the stem just above a set of leaves, and cut deeper down than you think. A general rule of thumb is twelve inches or more if the flower allows, especially for cut-and-come-again varieties like cosmos, zinnias, and dahlias. Usually, I determine where to cut a flower by feeling. I'll reach down the stem with my snips, looking for a set of leaves, and when the flower petals up at the top of the bloom gently rub against the inside of my elbow, I know it is time to start cutting.

After multiple snips and my hand full of flowers, I tuck the snips into my apron. Finally, I'll strip the flowers of any leaves that don't excite me for design purposes by wrapping my fingers around the stem and slowly pulling downward. Often I toss these leaves back into the garden bed, if the leaves and flowers look healthy, to act as organic matter in the soil. However, if it is very rainy, you'll want to be sure mold does not develop on these garden scraps, so scatter them with plenty of space in between.

Next, plop the blooms into a bucket of cold water. Bring the entire bucket indoors to a cool, dark space and let the blooms rest and rehydrate. I enjoy harvesting in the morning, with a cup of coffee, and creating with the flowers in the evening as the sun goes down. In the morning, my mind is often busy with the unwanted tasks of the day ahead. Harvesting blooms and being outdoors help to focus my attention on the beauty in front of me. In the evening, when my mind feels heavy and tired from life, arranging with flowers activates a space of creativity we often don't afford ourselves in adulthood. This combination of time in nature in the early morning sun and activation of the more creative side of the brain at night seems to be the perfect antidote to our often-chaotic lifestyle.

When to Harvest

There are some basic guidelines for the proper timing to cut a given flower. Many of these recommendations come from experienced flower farmers who sell to the general public or florists. Considerations such as storage and vase life are important when bringing blooms to market. However, for the average home gardener, you are given the freedom to harvest as you see fit. I find it great fun to snip a single flower variety over multiple mornings and take a moment each day to compare the blooms. This slow observation helps you deeply understand each flower you add to your garden over the years. However, in my early days of growing cut flowers, I found it very useful to have a clear suggestion for when to cut a flower. Following is a list of the flowers I grow and the most commonly suggested times to harvest each.

BULBS

The first of the flowers to bloom in the spring, these come on in waves of color and last longer in the vase when cut far earlier than one would anticipate!

- » **Daffodils**: Harvest when the flower looks like a thin pencil that has colored up and is just beginning to open. Sap will drip out of the stems after harvesting and is toxic, so be sure no one drinks the vase water. To arrange with other flowers, let these blooms sit in a vase for one day, then rinse the stems with cool water before putting them in an arrangement.
- » **Hyacinth**: This incredibly fragrant bloom should be harvested when most of the tiny flowers have opened; vase life is only about five days, so make sure to pay close attention to it while you are lucky enough to have it!
- » **Tulips:** Harvest when flower petals have begun to show color but have yet to open.

PERENNIALS

Perennials flow throughout the growing season; those that bloom early can often be harvested later in the season here and there for foliage. Of course, one might consider leaving some perennials in the garden to enjoy for longer stretches of time outdoors. But as the purpose of a cutting garden is to harvest and enjoy indoors, consider the suggestions below for the longest vase life.

- » **Astrantia:** Harvest when the outer petals have unfurled; it can also be harvested and used as a dried flower.
- » **Bearded iris:** Harvest just before the bloom opens up, when the small, pencil-shaped tip of the flower petals begins to color.
- » **Black-eyed Susan:** When the petals have fully opened, snip the blooms and place directly in cool water, as they are prone to wilting.
- » **Coneflower:** Snip this bloom once all the petals have fully unfurled and opened. It will slowly put on new growth for most of high summer.
- » **Delphinium:** Harvest when the bottom half of the blooms have opened and cut down to the soil line, which will encourage a second flush before the season's end.

- **Feverfew:** Harvest when flower clusters are roughly three-quarters open.

- **Foxglove:** Harvest when roughly one-third of the blossoms have opened. Remember that this flower is poisonous, so keep it out of reach of pets and children.

- **Goldenrod:** When half of the tiny yellow flowers have peeled open, harvest. Vase life can extend up to ten days.

- **Hollyhock:** I find this bloom very difficult to cut and bring indoors as it is so very stunning in a garden. When I do cut a stem or two, I harvest when about one-third to half of the blooms have opened for maximum vase life.

- **Lady's-mantle:** Harvest when roughly a quarter of the flowers are open or, after the blooms pass, harvest for foliage from spring through to the first frost.

- **Lavender:** When harvesting to use as a cut flower, wait until all the flowers on the long spike have opened.

- **Peony:** Harvest when the petals are closed tight but squish between your fingertips when squeezed. On a hot, sunny day, the texture and look of peonies can change rapidly, so you should check them a few times a day. Peonies can be stored dry for weeks in the refrigerator; just remember to wrap them in a cloth and lay them on their sides.

- **Pincushion flower:** Harvest when the flower petals are brightly colored and almost fully opened. Or leave them in place and harvest the seed pods after the petals have dropped.

- **Solomon's seal:** I harvest this plant all season long, and can't say I follow any particular harvest suggestions. Early in the season, tiny white blossoms decorate the edge of each leaf. In autumn, after the flowers pass, the leaves turn golden.

- **Tickseed:** Harvest just as the flower petals begin to unfurl.

- **Yarrow:** Harvest when full color develops and little bits of yellow pollen in the bloom center are visible.

Harvesting Woody Perennials and Shrubs

It is important to ensure the stems of woody perennials are able to pull up water to increase vase life. To do this, first make sure you have a clean cut at the bottom of the stem, and then split the stem in half by sliding a floral knife approximately a half inch deep. Let the stem hydrate in a clean bucket of water in a dark, cool room for the day.

ANNUALS

A cut flower garden wouldn't be complete without a selection of annual flowers, many of which produce more stems than their perennial counterparts.

- » **Ageratum:** Harvest when half of the small blooms open and have a textural appearance; cut the stems low to encourage additional blooms. The second flush, later in the season, will have longer stems. Arrange blooms straight away, as they don't hold over well.

- » **Blanketflower:** Harvest when the petals are nearly open, or let the petals drop in the field and harvest just the seed pods.

- » **Calendula:** This bloom is sticky when harvested, so keep that in mind. Snip when the petals begin to unfurl.

- » **Cockscomb:** Harvest when the flower is fully formed. Cut deep as this bloom will put on growth all season. The lower you cut, the longer the subsequent stems will be.

- » **Cosmos:** Harvest when the petals are near to bursting open and appear to be pulling away from one another. Vase life is fleeting, especially if harvested when the petals are more open.

- » **Dahlia:** When the first few rows of petals open, it is time to harvest. Cut deep, about twelve to eighteen inches, just above a leaf node.

- **'Frosted Explosion' grass:** This beautiful grass can be tricky to harvest; if you wait too long, the grass pulls wide open and is prone to tumbling out of the stem when cut. Therefore, I suggest harvesting when the grass is still slightly green and not yet fully blown open. This is a wonderful cut-and-come-again, so experiment with your own harvesting timeline, as you will have lots of it.

- **Globe amaranth:** Cut this branching flower deeply to encourage more stems. Harvest when almost all of the flower petals have expanded and opened.

- **'Irish Poet' tassel flower:** Harvest when the orange tassel blooms are visible but still held tightly together; once they have completely splayed open, the vase life dwindles.

- **Lisianthus:** I tend to harvest this at different times in its growing cycle as a single stem can have multiple blooms. I'll harvest a few stems when the very first bud on the stem opens, and on the remaining stems remove the first bud so all the energy is concentrated toward the later blooms.

- **Nigella:** Harvest it as a fresh flower when the petals begin to color or harvest the pods after the flowers have passed.

- **Snapdragon:** Harvest when half of the blooms have opened. Will store upright in a refrigerator for about five days, with or without water.

- **Strawflower:** Harvest when the petals have opened just enough to reveal the colored center of the bloom. As petals will close up at night or when it is cloudy, consider harvesting in the bright morning sun.

- **Zinnia:** Harvest just before the petals have completely opened. Make sure the stem is firm by wiggling it back and forth between your fingertips. After harvesting, change the water frequently, as its stems are prone to rot.

The Importance of Remaining Present

In late spring, with the first succession of flowers and the peonies in full bloom, it is safe to say we are at the start of peak cut flower season. Blooms grow in leaps and bounds seemingly overnight, and the aphids and groundhogs descend. It is easy to

feel each morning as though you are being met with a wall of frustration. All your well-intentioned plans have most likely been foiled in some shape or form.

This is why I encourage you to return to the healing benefits of a garden and show gratitude for the space you have created, even if it may not be exactly what you had envisioned in your mind's eye many months ago. Perhaps this morning, instead of immediately turning a critical eye toward uncompleted tasks, just enter the garden.

Sounds simple enough, but for a busy gardener it can be incredibly challenging. Can you fully enjoy the way each bloom curves out of the earth without worrying it will fail in the next wave of unexpected heat? Can you sit down and remind yourself of the texture of grass or soil while moving only your fingertips?

One method to fully become present in your garden is called grounding. This is the practice of returning to the present moment, and most of us do this at times of stress without even noticing. How often have you felt overwhelmed and suddenly noticed, without intention, that you take a large breath with a sigh? It's as if your body is reminding you that you are here and still breathing. Grounding in the garden can be as simple as bringing attention to your breath, the way your stomach expands and contracts, or focusing on what you feel and hear at a given moment.

I find the practice of grounding incredibly peaceful and easy; the only trick is to make space for it in your busy life. I invite you to learn more about grounding below and give yourself room to experiment with bringing this practice to your gardening.

How to Practice Grounding in the Garden

» Come to a comfortable seated position and take two to three large, deep breaths.

» Bring awareness to the weight of your body resting on the earth. Feel your hip bones sink into the grass, and call attention to how your skin feels where it makes contact with the earth. Begin to sense if the air is still or if there is a breeze running over your arms and face.

» Take in a large, deep breath and become aware of the scent in your garden. Perhaps you have planted herbs or flowers that you have forgotten to notice and enjoy. Or perhaps all you can smell is the earth. Continue to take long, deep breaths through the nose, becoming curious about the smells in your space.

» Next, listen to the sounds in your garden. Birds are an obvious outdoor sound this time of year, but can you detect anything that is subtle? The sound of a branch or leaves shifting in the wind, your own breath?

» Finally, if you have an edible bloom or herb in your garden, go ahead and pick and eat something. Use all your senses to experience slowly eating and swallowing the food of your garden.

Once you have grounded yourself completely in your garden, let your body rest in a state of present-moment awareness. Thank yourself for creating such a unique space, and simply be.

Summer
Harvest & Design

> But in the latter weeks of June there comes a time when I can begin to take breath and rest a little from these difficult yet pleasant labors; an interval when I may take time to consider, a morning when I may seek the hammock in the shady piazza, and, looking across my happy flower beds, let the sweet day sink deep into my heart.
>
> —Celia Thaxter

In high summer, the days often follow a similar pattern in coastal Maine; gray morning fog, the midday burn, and blistering-hot summer sun. I do my best to rise early and avoid the inevitable heat. When entering the garden at first light, I weed alongside the robins, who are keen for a bit of overturned soil and the promise of an earthworm. I harvest twice a week, on Monday and Friday, and the amount of blooming between three days in high summer is astounding. The flowers appear only outnumbered by aphids.

There is less room for pause at this time of year. My hands are busy with pulling high weeds and snipping blooms. Striking the balance between enjoying the garden and feeling exhausted by it proves a delicate art form. In this busy season, don't forget to give yourself space to just be, walk the garden with new eyes, and try hard to avoid the weight of an endless to-do list. Also, go easy on any grand plans you imagined for your garden that somehow never came to fruition. A garden is a lifelong project; it will never be complete in just one season or even a decade. That is part of the beauty—no matter what happens in the soil, the hope of next year carries you through.

If you find something is not working, whether a single flower or an entire crop, be ruthless. Letting go is something many of us struggle with in our day-to-day life. I encourage you to attempt a promenade, which, in ballet terms, is a slow pivot of the body while standing on one foot. How many times in life have you been twisted in an opposing direction, perhaps over the course of weeks or years, waiting to fall? But somehow, before you know it, you've arrived, unharmed, with a different perspective. Can you dig up a dahlia that once existed as a sign of hope and beauty, but just now lays rotting in summer rain? Might you convince yourself to begin again, sowing a sunflower in its place? Can you move slowly enough in the high heat of summer to recognize the lessons unfolding before you in the soil?

These lessons present themselves often, as the garden is teeming with metaphors. Describing them in words often cheapens the beauty of the thing. The seed is planted, it bursts through the earth, it is beautiful, it dies, the end. One draws a big, wide obvious line between the cycle of a flower and the cycle of human life, and sometimes in a foolish, near-cringeworthy way.

And yet, when in my garden, all my senses are dialed into a poppy unfurling in the morning sun, or a sunflower blackening, and the metaphors move through my fingertips to my soul. To the soul of the thing, past the flesh and bone. It is as if I were an instrument, a tuning fork on a small piece of earth, the initial vibration now stirring inside me, seemingly sent out light-years before. I enter summer with eyes wide open while tending my garden, in my floral design work, and my daily life.

A Beginner's Guide to Floral Design

The hard physical labor we call upon when working in the garden is very different from the slow, thoughtful way in which we work with flowers. As you settle in and begin creating with natural elements from your garden, I encourage you to find a small workspace for floral arranging. Ideally, it should be close to a sink, have storage for vases and snips, and be suitable for a bit of a mess! This could be the corner of a countertop, the washing machine when it's not in use, a spot in the garage,

The Importance of Sustainable Floral Design

In recent years the extent of waste created by the floral industry, by way of floral foam and single-use plastic, has become a major concern. In particular, research by scientists such as Dr. Charlene Trestrail has revealed that microplastics in floral foam often leach into surrounding water systems, impacting aquatic animals. Dr. Trestrail is an expert adviser for the Sustainable Floristry Network (SFN), an organization dedicated to educating and certifying florists on sustainable design methods. Most recently, the SFN called for zero-waste floristry and encourages those working with flowers to reuse containers, consider compostable products, and work with vases that are shaped to hold flowers without the need for additional mechanics, such as tape or foam. As the planet's health is always on my mind as a gardener, incorporating the suggestions provided by the SFN while creating even the smallest bedside flower arrangement is hugely important.

outdoors on a picnic table, or (if you are incredibly lucky) an entire room. Tuck all your floral design supplies together in an old sewing basket or shelf, as some supplies can be sharp and dangerous for little hands. It also prevents you from being tempted to use the indoor supplies out in the dirt of the garden.

The tools you will need to start designing your own floral arrangements are:

» **Snips:** My only advice here is to buy one or two very nice pairs of snips and a handful of inexpensive ones. I *always* misplace my snips, usually not far from where I stand, but if you have multiple sets, you won't be wandering around awkwardly holding fragile blooms just looking for a cut.

» **Five-gallon bucket:** This is for collecting any snipped stems or leaves as you work, which can then be thrown into the compost pile when you are finished. You can also flip a bucket over on top of the table and use it to bring your arrangement to eye level.

» **Lazy Susan:** Place this on top of your table or bucket so you can rotate your design with ease; examining it from all angles is important if you are creating a centerpiece.

» **Floral knife:** I use mine to remove thorns and other little bits of imperfection on blooms. These are quite sharp, so exercise caution and extreme focus when handling.

» **Floral frog:** One of my favorite tools for designing, a floral frog sits in the bottom of a vase and looks to be a circle full of tiny thumbtacks. It will tightly hold each flower stem that you place in an arrangement, ensuring that even blooms set at an extreme angle will not move. Floral frogs have been in use since the fourteenth century, originating in Japan. It is highly likely you will stumble upon glass or vintage floral frogs at antique stores. Glass floral frogs have large holes, and thus aren't the best for arranging a collection of thin-stemmed flowers. I do find them useful for flowers with large stems, like hydrangeas and tulips. Many floral designers have returned to floral frogs, as they are sustainable and reusable.

» **Chicken wire:** Wonderful for use with large vases or heavy flowers, chicken wire is cut and then rolled to shape and fit into your vase. Consider using a vase

that has a bit of an edge so that you can tuck your wire beneath it for stability. Be careful not to scrunch up the chicken wire too much, as this will make it difficult to slide in bigger stems without splitting them. Remove and reuse!

- » **Twine:** I make sure to have jute twine and metal twine on hand for various floral projects. Jute twine gives a lovely organic look to floral bouquets, and metal twine works well for projects such as wreaths or floral crowns that require a bit more hold.

- » **Floral clay:** This is used on the bottom of a floral frog to hold it in place in the vase.

- » **Kraft paper wrap:** Useful to fancy up a hand-tied bouquet. Simply cut a square, fold it into a triangle, wrap it around your blooms, and tie with twine or ribbon.

- » **Ribbon:** Always have three or four varieties on hand. My favorites are those that have been created using natural dyes. A single ribbon adds so much to a bouquet.

> The glasses (thirty-two in all) themselves are beautiful; nearly all are white, clear and pure, with a few pale green and paler rose and delicate blue, one or two of richer pink, all brilliantly clear and filled with absolutely colorless water, through which the stems show their slender green lengths.
>
> —Celia Thaxter

Sourcing Your Vessels

My collection of antique vases, sourced over decades, has slowly taken over every shelf in our farmhouse. The reason I am drawn to antique vases has much to do with their functionality, as they were created before things such as foam or floral tape. Ceramicists designed these vases to hold flowers just so, without additional mechanics. For example, a tulip vase was crafted to support the delicate bend of this flower as it ages and droops downward. Oftentimes an antique vase is etched or painted with the flower it is designed to hold. The beauty of a vessel highlighting the life cycle and shape of a flower, as opposed to shoving it into foam and forcing it to stand up straight, is never lost on me. Watching a single arrangement shift, bend, and ultimately drop its petals is an incredible form of meditation.

Some types of vessels you may encounter include:

- **Bowls:** These are fantastic to design with when you have lots of short-stemmed or airy flowers that don't need a ton of support. Most often I used a single floral frog in the bottom of a bowl.

- **Bud vases:** Perfect for a flower that needs to sing alone. Often when I put together an arrangement, I'll find one or two flowers that simply need their own space. Usually this is a bloom with an interesting stem shape or a flower bud that is so tiny it becomes lost in a larger arrangement.

- **Classic bouquet vase:** These are wonderful for long-stemmed wildflowers. I fill mine with foraged Queen Anne's lace and goldenrod all summer.

- **Floral brick:** Popularized in the eighteenth century, this is often used to display cut flowers from the home garden.

- **Footed vase:** This type of vase is a bit more formal. Consider using one if you are having a dinner party or your most decadent focal flower has bloomed.

- **Glass vase:** Many flowers' stems are so gorgeous that it is a shame to hide them behind a solid ceramic vase. Consider purchasing these vases in a variety of shapes and heights so that you might have one ideal for smaller flowers, such as a snowdrop, or bigger blooms like a long-stemmed dahlia.

- **Narrow-top vase:** A carefree vessel to use after you have wandered the garden and created a small hand-tied bouquet; just plop the stems in the water.

- **Simple pitcher:** Perfect for a casual arrangement on the kitchen table, my pitcher functions as a vessel to hold a handful of flowers cut in the early morning that I don't have time to sit down and more officially design.

Setting the Stage

Each time I design with flowers, there is a bit of poetic ceremony that unfolds. I tend to arrange my flowers on our dining room table, which has never truly functioned as a formal setting for meals; it is more of a workspace for forgotten Lego projects and science experiments. To begin, I slowly take my supplies out of an old

wls

Bud vases

ssic bouquet vase

Floral brick

oted vase

Glass vase

rrow-top vase

Simple pitcher

sewing basket, inspect them for any signs of rust or dullness, and place them neatly on the table. Next, I fill my favorite indoor pitcher with cool water and, if I'm lucky enough to be working with a bucket full of blooms, I pour water into various vases or mason jars. Gently pulling each flower from the large bucket, I organize the blooms by likeness or flower type, placing them into individual jars.

    Taking into consideration the height of the flowers, or stem length, I select a vessel that will allow for the tallest blooms to float above an arrangement and the smallest stems to hover just over the lip. After finding the perfect vase, one must consider how the flowers will hold in your arrangement, which is often dependent on the weight of the bloom. I gently lift each flower from its jar, balancing the petals between my fingertips to get a feel for its stability. Flowers that are heavy, such as dahlias or peonies, may need not only a floral frog but a layer of chicken wire and perhaps some strong woody stems for security. Selecting the correct vase and mechanics for your arrangement is like building the foundation of a home. When I have rushed the process, inevitably, the entire arrangement tips over within the hour and I'm left with a pile of snapped stems. So, think about vessel selection, and slowly secure your wire and floral frog. Once your mechanics are firmly in place, fill your vessel with water.

Before snipping the first flower, close your eyes, take a deep breath, and arrive. There are no rules when playing with the garden flowers you have grown. Arranging flowers is more about the act of handling each petal with your attention fully engaged, not the final product. I must admit that I believe garden-grown flowers, even hand-tossed in an old vase, look stunning. That being said, most of us desire some hints when arranging blooms.

Fourteen Tips for Floral Design

» Practice nonjudgment. Often, we cannot get out of our own way. Our critical mind appears uninvited during our creative process. Consider turning on some music and make sure you are given space to be alone while working with blooms. If you find your mind creeping toward negative self-talk, calmly request this narrative wander off for a bit, and move on.

» Take time to admire each flower. Closely examine the curve of its stem, twirl the bloom between your fingertips, viewing it from different angles. Perhaps gently pull open a few petals by warming and rubbing them between your thumb and index finger.

» Set up so that you are at eye level with your arrangement. A small coffee table or bucket flipped over and placed upon a table can give you the little boost needed.

» Use a lazy Susan to spin your design so you can easily examine it from all sides.

» Traditionally, floral arrangements should be one and a half to twice as high and wide as your vase. However, there are no absolutes; many new style arrangements incorporate an asymmetric design, whose shape makes it difficult to discern measurements like those just described.

» Consider working in odd numbers, ensuring you have three, five, or seven of a particular flower.

» Design so that the same flower or flower type is at different heights and depths throughout your design.

» Use focal flowers sparingly; too many and they lose their magic.

» Don't be afraid of empty space. This might be my greatest struggle; most of my arrangements end with the pulling of ten or more flower stems that were stuffed into wide beautiful bits of air between decadent blooms.

» Work with the natural curve of flowers. Stems that achingly arch in all directions are unique to the home garden. Consider the strength required for a given bloom, pulling from its root, to twist and reach for the sun. If this sentiment moves you, perhaps place this single flower in a vase. Less is often more, and don't underestimate the ability of just one flower to tell a meaningful story.

» Cover the lip of the vase with greenery or low-lying flowers. I enjoy using a trailing piece of greenery from nasturtium or hyacinth bean to cascade from the vase's edge to the tabletop.

» Some individuals start with greenery in their design but I prefer to start with focal flowers. I would encourage you to play around with both options. Shrubs and woody perennials placed in an arrangement first can be a wonderful source of structure. Using focal flowers first ensures the most interesting elements of an arrangement are placed just so, not as an afterthought.

» Soften harsh and aggressively straight stem lines by covering them with light and airy filler flowers.

» If you hit a wall, stand back, walk away, perhaps stroll the garden for inspiration, and return. When designing flower arrangements from the home garden, there is no rush.

The Power of Designing in Darkness

When playing with flowers, I often struggle with quieting my mind and inner critic. I twist each stem between my fingers over and over again, searching for the correct angle or location to cut the bloom. Sometimes, I overthink things so much that I just stand, staring into a bucket of freshly cut flowers that are unmovable.

On one such evening, with a lush harvest of daffodils staring up at me, I put down my snips, turned off the lights, and fell quiet. There under the moonlight, the daffodils glowed. My mind became suddenly aware of lines and details in the darkened petals I hadn't noticed before. And not dissimilar to a field lighting up with fireflies, my thought process came alive, flooded with memories of exploration under the summer sky and all the secret paths in the woods I'd traveled. Places that some dear friend brought me to as I questioned, "Where are we going?" and they hollered, "You'll know when we get there."

I am lost in memory: down the Brooks' path to a York River filled with bioluminescence, around Crescent Beach to Seapoint, through the woods and over the trolley tracks to Brave Boat Harbor, to Cow Beach before the Cliff Path was chained off in York Harbor, a foundation of a home from a hundred years ago, and the long-forgotten amphitheater off Walt Kuhn Road in Cape Neddick.

You have the same type of paths in your memory bank—at least I hope you do. They are the places with brambles underfoot, marked by a break in the wildflower line, where stems gently bend in half and rub your calf as you pass. The ones that, when you descend into them, the air temperature drops a good ten degrees, and you wonder if you've made it to Narnia. Hidden by the pine trees from the tourists and the traffic lights, tucked into a forgotten space. A look behind reminds you of the chaos that is, and a look ahead reminds you of the peace that was before the noise.

I couldn't see all this with the lights on, but when all these places and people washed over me, my hands began to move from memory. How many times had I gently rubbed my fingertips atop the varied heights of wildflowers on a side road? I didn't have to overthink the spacing and dimensions; it was a basic knowledge that lived inside of me. When you design with local garden-grown flowers, you almost unwillingly open a portal of nostalgia. By turning off the lights and quieting my mind, I was able to let my hands and the blooms tell the story of place.

Will everyone understand why I made a floral arrangement in the dark, complete with snapped stems and wild raspberry branches? Of course not, but that is the luxury of creating with my own flowers. Often, it is more about my process than the finished product, and the same goes for you. The garden acts as a source of truth about things we've forgotten: that the world can still be beautiful and that being outdoors is good for your soul.

Tending the Summer Garden

In summer, when walking the garden and honing your skill of observation, you will likely notice that the garden seems … tired. High heat causes leaves to droop, and various insects and wildlife descend on petals and stems, searching for nourishment. In spite of it all, the flowers in summer hum with life. Cut-and-come-again varieties produce stem after stem at an almost unfathomable rate, and the last of your seedlings take off with gusto. During this season, it is so very important to tend to your garden with great care. Reducing stress on plants and enhancing soil biology are necessary skills for a gardener to develop, ensuring blooms carry on until the very last frost.

Dealing with Pests in the Garden

Harmony, specifically with bugs and wild creatures, might seem a laughable subject in high summer. Often, you will see the row garden in our backyard outlined with cayenne pepper (I truly can't say whether this works) and find me in the evenings pulling slugs one by one from the zinnias. Nothing about these activities feels harmonious. But perhaps I need to expand my perspective.

    I do believe we need to share the earth and that my zinnias are not exclusively mine. The more I let go and relax around the inevitable groundhog assault, the more comical it becomes. And I hold deep respect for the gardeners who have returned their space to nature by creating wildlife habitats with their own two hands. However, as I grow flowers to bring to market, it is also important that I strike a balance between acceptance and financial ruin.

    When I think of the term "pest," I'm reminded of an enormous aphid outbreak last spring. My beloved lupines were covered in hundreds of green bugs that made my skin crawl. After spraying the aphids with a hose a few times, only to realize this

SUMMER | 143

forcefully pulled the beautiful purple petals from each stem, I gave up. Within a few days, birds descended upon the lupines and, naturally, these "pests" were gone.

I would never suggest you let slugs devour your precious flowers, but I will say that observing nature and expanding your knowledge on why a certain bug has descended on your blooms can help you grow alongside the insects that call your garden home. Please know I'm not some enlightened gardener floating around and smiling at stems chewed to the ground. I absolutely cried when a deer ate the beautiful hollyhock my children gifted me in early spring. But I will say I've perfected the art of a big cry followed by a big laugh.

Following are some ways I've tried to reduce damage to the flower garden. Many of these methods are inspired by what's known as Integrated Pest Management (IPM), an environmentally friendly means for controlling pests that focuses on observation, encourages education on the life cycle of pests, and discourages the use of chemical pesticides.

GROUNDHOGS

The only way to completely resolve your groundhog problem is to put up fencing that extends over and under the soil. Digging a trench and laying fencing beneath the soil will prevent these animals from burrowing into your garden.

DEER

Similar to groundhogs, it is all in the fencing to fully protect your flowers. Deer-Busters® makes affordable, easy-to-install deer fencing. Here at the farmhouse, I keep more valuable crops, like poppies or other focal flowers, in an area fenced with chicken wire. In the larger open garden, I have worked hard over the years to cultivate deer-resistant perennials with a small mix of annuals. These include dahlias, foxglove, coneflower, herbs, yarrow, and peonies.

SLUGS

Notorious for attacking garden favorites like dahlias, slugs constantly chew large holes in my flowers. I have found that Sluggo®, which the Organic Materials Review Institute has listed for use in organic gardening, when applied at night every two weeks or so, can considerably reduce the slug population. I also remove them by hand when I notice them at twilight or in the early morning.

APHIDS

These tiny green bugs will absolutely cover your flowers. Simply spraying water on these critters each day and knocking them off the blooms are the best solutions. In most cases they cause no damage to the plant but are unsettling if you are cutting flowers to bring into your home.

JAPANESE BEETLES

If you notice the leaves of your plants have a skeletonized appearance, Japanese beetles are the likely culprit. I recommend using soap and water on your plants and removing the beetles by hand. If you walk your garden and observe the petals and stems of your blooms every day, noticing a pest outbreak and addressing it while it is small and manageable is possible.

Nourishing Your Soil Matters

As you might expect, I believe the best practice for any gardener seeking to nourish their soil is by applying all-natural amendments. Luckily, there are plenty of organic

Three Tips for Early Pest Prevention

» **Increase your plant health:** Most pests attack plants that are already weakened in some way. So it is important to keep up on your weekly fertilizing.

» **Practice companion planting:** Planting herbs and flowers that deter pests next to high-value crops is another method of pest control. Herbs such as dill and basil and flowers such as nasturtium, marigold, and feverfew are known to keep bugs away. I have had success planting feverfew around my lisianthus and other high-value crops over the years.

» **Rotate your crops:** Just as it sounds, this practice involves moving crops to different locations each season. This is why keeping notes, and reviewing your winter planting plans, is important for years to come. By moving crops to different locations, pests that overwinter as larvae will be surprised to find their favorite flower is no longer available when they wake up in spring.

options when it comes to soil improvement. Some are simple, premixed, and ready to use, and others involve a bit of experimentation with items found around the house. I will review some of my favorites, from least to most involved.

FISH AND SEAWEED FERTILIZER

I will warn you, these smell quite terrible but they are my go-to ready-to-use fertilizers. I use them once a week in the growing season, but every other week will work if you aren't harvesting lots of blooms. Be sure to read the fine print on the back and measure properly; a little goes a long way. Most fertilizers suggest one ounce per gallon of water. I try to avoid fertilizing on very hot days as the smell can be intense under the bright sun.

MOLASSES

Believe it or not, one to two tablespoons of molasses mixed into a one-gallon bucket of water is a wonderful foliar spray for your flowers once they are growing. I had my doubts about this method, but last summer when my dahlias looked incredibly tired and overheated, an application of molasses in the morning perked them up by midafternoon. Molasses gives your blooms a sugar boost and provides vital nutrients and microorganisms.

HOME WORM FARM

Consider a small worm composter tray. It is so valuable and easy to tuck into a corner in your garden. There are lots of different varieties of home worm farms to choose from but the general design is essentially the same. A roughly two-foot-high-by-two-foot-wide tray system is filled with soil, kitchen scraps, and purchased worms. The worm waste collects at the bottom of the tray system and, using a small spigot, you can pour worm castings into a bucket. Once you have retrieved the worm castings, mix them with water and apply the mixture to your garden.

SEAWEED

As I am lucky enough to live by the sea, from time to time I'll bring home a bucket of seaweed and use it in my garden. In the fall I place it around precious perennials to insulate them from the cold, harsh winter. In spring, I enjoy broad-forking it into the soil as I prepare beds. Seaweed can also be made into a tea by letting it sit in a bucket of rainwater for two to four weeks. Add a small amount of this liquid to your watering can a few times a week. Or you can dry seaweed, crush it, and sprinkle around the base of your plants in high summer when they look tired.

Beginners Can Save Seeds!

The act of saving a seed, to me, is a sacred one. And it is not just for the more experienced gardeners! A handful of seeds can produce an abundance of food or flowers and is very

Lobster Shell Extract: A Recipe

Korean natural farming is a practice that encourages soil improvement by introducing bacteria and fungi native to your growing space and applying these inputs at the proper time. If this topic interests you, I highly recommend checking out Nigel Palmer's book, *The Regenerative Grower's Guide to Garden Amendments*. Using the recipes in this book, I created my own lobster shell extract, which I combined with molasses and worm castings and fed to my flowers throughout the growing season. They were so very happy! This recipe is perfect for folks living on the coasts with access to lobster shells, but also for anyone who finds themselves with lobster shells after a celebratory dinner.

To make the lobster shell extract:

» Convince a local lobsterman to drop off lobster shells or have an old-fashioned lobster bake and save the shells. Clean and then cook the shells in the oven to remove any remaining lobster meat. Remove from the oven and crush the shells finely with a mallet.

» Place the shells in a large jar and add apple cider vinegar, following a ten-to-one ratio, with one part being the lobster shells.

» Place cheesecloth over the top of the jar and secure it with an elastic. Place the jar in the back of a dark closet for seven to ten days.

» Strain the liquid out of the jar, shake out the lobster shells, and place the liquid back in the jar.

» I like to mix approximately one teaspoon of this mixture with a gallon of water and apply it to my soil.

If this feels like a lot of work, not to worry! Remember you can always order a premixed fertilizer that is ready to go in your watering can as soon as it arrives.

empowering, summoning hope for the future. The cultivation of seeds has an important history: it allowed humans to stay in place. Historically, the role of seed saving and tending to the garden was tasked to women, who acutely understood the role of seeds for survival. In 1862, Dakota women sewed seeds into their skirts in preparation for a 150-mile forced walk from Lower Sioux Reservation to a prison camp at Fort Snelling. The significance of a single seed should never be overlooked. Without it, we wouldn't have the garden at all.

When you look at a flower, dried and browned on its stem, let it stay awhile before cutting it down. Consider saving some seeds, storing them in a paper bag, and replanting next spring. Seeds harvested from your own garden are unique to your soil, weather, and pollinators. It's a great feeling to revel in the magic of a handful of seeds displaced from a single bloom that was perhaps planted by your own hands months before.

Not all flowers make great candidates for seed saving, however, and for this section, I will need to call on a little science. You want to save seeds that are open-pollinated, not hybrid. If you harvest seeds from a hybrid flower, you might get an odd collection of traits on a bloom the following spring. This is not the end of the world, of course; it is just a lesson that a hybrid will not create an exact replica of its parent bloom.

How to Save Seeds

» Allow the flower to stay in your garden until all the petals have fallen off and the seed pod has browned.

» Cut the seed pod off the flower, place it on a tray, and bring it to a location with good lighting. It may be difficult to tell what is a seed and what is simply plant matter.

» Gently pry open the seed pod, or shake, releasing the seeds from the flower.

» Allow seeds to dry for one week in a cool, dark, well-ventilated space.

» Place seeds in a paper bag and store them out of sunlight over the winter.

Also, be careful of taking seeds from garden beds where you grew varieties of a single bloom close together. Once again, flowers produced from these seeds may be very different from the ones you are expecting. This may be a fun experiment as a home gardener, but if you are growing flowers to bring to market, this unpredictable outcome may not be ideal.

Some of my favorite varieties of seeds to save are sunflower, hyacinth bean, nigella, zinnia, calendula, black-eyed Susan, coneflower, and nasturtium. I tend to look for easy-to-harvest seeds with obvious seed pods.

Considering Native Flowers in Your Cutting Garden

As I look out my window just now, dotting the tree line around the perennial garden are collections of wild geranium, goldenrod, and New England aster. Native flowers are beautiful and very important for attracting beneficial insects and increasing biodiversity in the garden but are often overlooked when it comes to floral arranging and cut flower gardens. I have heard complaints like their vase life is poor, or they look like roadside weeds. But I assure you, there are plenty of native flowers that will look beautiful in your landscape and function well in arrangements.

Native plants are important to add to your cutting garden as they benefit pollinators. Because of this, I harvest only two-thirds of these blooms and leave the rest for the bees. I must warn you, when you start growing natives, you will notice

an influx of adorable bees sleeping upon your flower petals, making it very difficult to cut the stem and harvest. And you will be astounded by how many pollinators are attracted to your garden. Nothing is more peaceful than the vibrational hum of a collection of bees or an abundance of low-flying butterflies that dip and weave between your flower beds.

    A note about natives: just like all cut flowers, you must be mindful of when you harvest them. New England aster can have an acceptable vase life if you harvest before all the petals are open. Too often, we harvest a native bloom after its prime.

This is because they tend to be discovered, not cultivated, in our gardens, and harvested when we stumble upon a given bloom in a parking lot or vacant field. You will find it is difficult to find resources on the best vase life for native cut flowers. Therefore, I encourage you to experiment! Harvest at various stages in a flower's life cycle and record your observations.

In this section I will offer a list of my favorite native blooms to use as cut flowers. Many native blooms grown from seed should be placed in the soil in late fall and early winter as they require a bit of cold to germinate. So, always remember to do your research on the seed-starting timeline. (The Wild Seed Project and the Lady Bird Johnson Wildflower Center are both wonderful resources for information about native plants.) If you aren't going to start from seeds, be sure to source your plants from a reputable native plant nursery near you.

Finally, I just want to say one thing about natives versus cultivars. Many beautiful blooms labeled native at your local nursery are actually cultivated versions of the original native flowers; these are known as nativars. These plants have been altered in appearance for desirable traits such as color or petal shape. Many gardeners take issue with the use of nativars in a garden, especially when they are showcased in an area of a nursery and described as native plants.

While I agree with the need for plants to be correctly labeled, the truth is that very few nurseries carry *true* native plants. This makes it difficult for the average home gardener to curate a native flower garden. However, not all nativars are as bad as one might think. In 2016, Annie White of the University of Vermont completed a study that revealed nativars such as *Asclepias tuberosa* 'Hello Yellow' and *Monarda fistulosa* 'Claire Grace' attracted the same amount of pollinators as their true native counterparts. In fact, the nativar *Veronicastrum virginicum* 'Lavendelturm' attracted more pollinators over the season than the original native, due to its long bloom window. So, while I fully support the notion that as a general rule native plants are best for bees and butterflies, I also believe that if a nativar excites you, especially one that closely resembles the original native, consider planting it in your garden and observing how pollinators react.

Top Native Plants for Cutting

- **Black-eyed Susan** (*Rudbeckia hirta*): I'd argue that this is one of the most popular native blooms found in home gardens. It is rugged and deer-resistant and produces vibrant color for much of the summer season.

- **Coneflower** (*Echinacea purpurea*): This beautiful bloom comes in many varieties and cultivars but the true native is still my favorite. The blooms are a delicate light purple. Don't be afraid to snip as this flower will put on blooms all season.

- **Joe-pye weed** (*Eutrochium* spp.): This plant can grow up to seven feet tall. Its muted purple flowers and large, structured leaflets make it a lovely bit of foliage. I think it looks striking planted next to goldenrod; the colors make a wonderful contrast to one another.

- **Northern sea oats** (*Chasmanthium latifolium*): This beautiful native grass is not a huge producer, but the small, delicate golden seed pods are wonderful in dried flower arrangements.

- **Phlox paniculata:** This has lovely pink blooms that often grow so large I simply cut a few and place them in an urn. This bloom is deer-resistant and likes moist soil.

- **Wild bergamot** (*Monarda* spp.): Also known as bee balm, this blossom reminds me of someone who has woken up from a late night of partying. It has beautiful light purple petals and a hard woody stem. This is a top favorite of bees.

- **Yarrow** (*Achillea millefolium*): Common yarrow is a wonderful flower to forage, found on roadsides for most of the summer. The native yarrow is white, while cultivars come in various colors.

Taking Time to Connect

Often the act of gardening or floral arrangement is a solo pursuit. These interests provide a space to clear your mind and tuck yourself away from the chaos of the world. And while it is incredibly romantic to have a simple kitchen table dripping with freshly cut flowers, there comes a point in the season where one feels a bit self-indulgent. To be in the presence of such beauty and not share it with others can feel wrong. It is important as flower growers to consider ways to connect with our community through our harvest.

Give Your Flowers Away

The unexpected gift of flowers to a dear friend or total stranger can change the course of one's day. But, I must warn you, the act of giving blooms is highly addictive. In 2021, I was chosen to be an ambassador for the Growing Kindness Project, founded by Deanna Kitchen. The purpose of the project is simple: grow and give away flowers. In early spring, I was sent ten dahlia tubers, and hundreds of flower stems were harvested from three simple raised beds we installed in the front yard just a few months later. Every stem was given away to community members, with an emphasis placed on health-care workers and educators. And in the fall, the tubers I dug up were shared with others interested in continuing the project.

The experience of giving away flowers had a profound effect on me. The structure of people's faces changed; they softened or melted in front of my eyes. Most people didn't understand why they should have flowers and questioned if they truly deserved them. People who had worked tirelessly for the benefit of others during the pandemic could not grasp why they were given a handful of dahlias. This taught me never to underestimate the power of a homegrown flower.

A flower cultivated by your own two hands has an intimate and unique effect on others, very different from the response to a grocery-store bouquet. So, when you have extra blooms, consider placing a small collection in a vase or upcycled jar and sharing them with a stranger. Also, if you are growing vegetables and have some to spare, contact your local food pantry or state master gardener program. Many of these organizations are more than happy to accept produce donations from local gardeners.

Open the Garden Gate

The use of seeds for agricultural purposes began during the Neolithic revolution. Once food could be cultivated, the need for hunting and gathering ceased and we began to stay in place. "Garden" is derived from the Middle French word *gardin*, which implies a section of land that is enclosed and cultivated. This is not a poorly veiled attempt at suggesting how important a fence is, but merely a reminder that gardens were not always meant to be pretty, and justly so. They began as a functional space to grow food and sustain life.

Fast forward to 150 BC Rome, and you will begin to notice ornamental gardens in the homes of the wealthy. Walking and pontificating among narcissus and oleander became a desired activity. In Athens, Aristotle famously gave lectures in the groves of the Lyceum. In early colonial homes in the United States, the traditional kitchen or agricultural garden was hidden in the back of the property, while the ornamental garden sat in the front or a side yard on view from the nicest room of the house. Gardens have often been a place of connection, a stimulating environment to discuss politics or consider scientific discovery.

Today, a summer walk in someone's garden is an intimate thing. Often spur of the moment, unrehearsed, and humble, you see a gentler side of a neighbor or acquaintance. These walkabouts tend to start with an explanation of unrealized dreams: "I was hoping to put a clematis there, but I didn't tend to the soil in time, and now I've half given up." Often there are apologies: "Oh my, I'm covered in dirt, excuse my nails." But then, as if by magic, you are in front of the most beautiful spot in the garden. It might be a simple flower, or a row of blooms, neatly weeded. Whatever this crown jewel is, your tour guide is pleased you've seen it. We can go days, weeks even, without a single soul seeing our garden.

In 2023, the surgeon general of the United States released an advisory about the amount of loneliness in this country. Today, half of the adults in the US admit to experiencing loneliness. This is precisely why I'd encourage you to share your garden with someone—perhaps the neighbor across the street whose lilies you've coveted for years or the colleague at work with whom you compare growing notes over lunch. Consider opening the gate to your garden this season and showing someone what you have created. You might inspire a new gardener to plant a zinnia, sunflower, or calendula.

Florals in the Kitchen

I'll be completely honest—cooking is not my strong suit. However, I'm always happy to provide that floral pop needed to fancy up a dessert, cup of tea, or cocktail. There is something so very glamorous in the way calendula petals float in your best champagne glass. And my children love adding flower petals to the top of a cake, made for no special occasion on any given Tuesday. Flowers can make the simplest item, like an ice cube, seem like a bit of poetry. In that spirit I will share some wonderful edible flowers, and ideas for using them simply in your day-to-day life.

- **Calendula:** I enjoy drying these bright orange petals to make a floral confetti of sorts, sprinkling the curled petals on cookies, cakes, and beverages, especially in winter when the vibrant color almost warms you simply by looking at it.

- **Chamomile** (*Chamaemelum nobile*)**:** One of the most popular medical herbs, commonly used in tea, but also a delicate little flower to use as a garnish on a dessert or as a salad topper.

- **Lavender:** A most heavenly scent, lavender is often used when someone needs a bit of calm. I enjoy it best stitched in a sachet to rest on my eyes just after turning out the light for bed. In the kitchen, it is so lovely as a simple syrup, traditionally used for alcoholic drinks, but of course you could add it to seltzer water or tonic. To make simple syrup, place one cup of water and one cup of sugar with at least one tablespoon of lavender blooms in a small saucepan. Turn the heat on high and stir until the sugar has completely dissolved. Remove the saucepan from the heat and let the syrup sit for a half hour or more. Strain the lavender out using cheesecloth and place the syrup in a pint mason jar, labeled with date.

- **Lemon balm** (*Melissa officinalis*)**:** I adore this herb in a bouquet; its smell is so crisp and refreshing. But I've also been known to tuck a sprig of it next to a scoop of vanilla ice cream on a hot summer night.

- **Nasturtium** (*Nasturtium* spp.)**:** This beautiful climber is one of the most commonly used edible flowers to spruce up a plate of food. I like to cut the orange bloom and add it to a large ice-cube tray with water. This flower ice cube is so cheerful in a simple cup of water or for guests in a signature cocktail.

- **Red clover** (*Trifolium pratense*)**:** Often found growing wild in lawns and along roadsides, this perennial is known to help with soil erosion and is nitrogen-fixing. It is also the very first flower I ever tasted. My favorite way to enjoy red clover is to pick it, pull out a tiny purple blossom, and suck on the bottom. The sweetest bit of nectar is released.

- **Yarrow:** Cultivate it in your garden or pull it from the roadside where it runs wild (for medicinal purposes, the white wild variety is best). Yarrow can be used like many of the other flowers listed here, as a garnish, to top a salad or cake, or dried and used all year round.

Considering Natural Dyes

By now, you probably recognize that I believe the garden is a wonderful place to unwind and find stillness. This does not make me unique; many people report decreased stress levels and an increased sense of general well-being after working or spending time in their garden. But what happens when your winters are long, and the garden is covered in snow?

Over the years, I've searched for ways to use material from my garden or the outdoors to extend my connection with the soil. One of the ways I have accomplished this is by natural dyeing fabrics. I find benefits similar to those I get from gardening when I spend an afternoon handling organic material, focusing on a simple task (such as boiling water) and creating something beautiful. I would never describe myself as a crafty person; fine motor activities are not my strength! This is why gardening and natural dyeing are a good fit for my skill level.

Similar to gardening, there is some experimenting and science that goes along with the process. Certain elements, like willow tree leaves, result in a more vibrant

color when you leave them soaking for over a week. Using an iron mixture to treat fabric after dyeing causes a total shift in color; marigold-dyed fabric turns from deep yellow to green. There can be lots of play in the dyeing process, which most of us do not afford ourselves regularly.

If you have been curious about working with natural dyes, here is a quick beginner's guide. I have had success dyeing ribbon, dinner napkins, and silk scarves. I highly encourage you to try it while you still have some flowers in bloom, but also know that dyeing can be accomplished all year. Things like acorns, pinecones, and lichen are always available in the winter months. The seasonality of natural dyes is part of the fun.

How to Naturally Dye Fabric

Many of the items you need for natural dyes are common kitchen supplies and can be sourced affordably secondhand. Designate certain pots, tongs, and spoons for natural dyeing and don't interchange them with items you use for cooking food! Wear gloves when preparing and treating fabric.

- » Large pot
- » Wooden spoons
- » Tongs
- » Heavy-duty safety gloves
- » Mordant (this is a substance that fixes the dye to the fabric)
- » Strainer
- » Measuring spoons
- » Kitchen scale

Extra fun:
- » Soy milk
- » Clamps
- » Rubber bands

STEPS

- » **Select a natural fabric:** You must select a natural fabric as natural dyes will only adhere to natural fabrics. Consider either 100 percent organic cotton or linen. I enjoy dyeing bits of ribbon to use in seasonal wreaths or to wrap in my hair. Soft cotton squares can be dyed and used as dinner napkins.

- » **Weigh your fabric:** Once you have selected your dyeing material, weigh the fabric on a kitchen scale and record your result.

» **Score the fabric:** Bring a pot of water to a boil, put in your fabric, and let it boil for ten minutes. Turn the heat down to low and let it simmer for one hour. After this, remove the fabric from the water and toss it into your washer on a low-temperature gentle cycle with your favorite natural laundry detergent (traditional detergents often have bleach, which will change the color of your fabric).

» **Pretreat the fabric:** While the fabric is washing, empty the pot, fill it again with water, bring it to a simmer, then add the mordant, which will prepare the fabric to receive and keep the color over time. I use alum as a mordant, which is easy to purchase from various sources online. A general rule is the alum weight should be equal to 15 percent of the weight of your fabric. Rarely do I use more than a teaspoon. Place the clean, wet fabric in the pot and let it simmer for approximately one hour. Be sure to use a wooden spoon to move the fabric around every so often. You want the fabric to be able to move in the water so that it is fully exposed to the mordant. Turn the heat off and let the fabric cool in the pot; you can remove it after an hour or leave it soaking overnight.

» **Create a dye vat:** Fill a separate pot with water, add your dyeing medium (flowers, berries, etc.), and bring this to a boil. For the best color, the weight of your dyeing medium should be roughly equal to the weight of your fabric. You can always play around with the amount of dyeing medium to see how it impacts the intensity of color. In most cases, natural dyes are subtle, so the more plant medium, the better the color.

» **Dye the fabric:** After the water has come to a boil, turn it to low and, when the boil has slowed, add the pretreated fabric. Often I pull the wet fabric with tongs from the mordant pot and toss it directly into the dye pot. Leave the fabric in the dye pot until the desired color is achieved. This could take as little as twenty minutes if you are using certain berries and up to a few days for something like willow tree leaves. For fabrics that need a longer soak, turn the heat off completely after an hour and periodically stir the fabric.

Once you have practiced using natural dyes, you might consider wrapping your fabric in rubber bands or clamping it to make polka dots. Also, applying soy milk with a paintbrush to pretreated fabric allows you to create shapes or lines on the fabric of a slightly different tone of color after the dyeing process is complete. Just be sure that the soy milk has dried before putting the fabric into the dye vat.

A few notes: Dye liquid is difficult to store, and often the color tires after you have dyed a few items. I recommend using as much dye as possible when it is freshly made. Be sure the window is open and a fan is on to ensure good airflow during this process. Once I started dyeing fabric more regularly, I purchased a small two-burner system and set it up outside.

Slowing Down in the Garden

Gardening is a wonderful way to unwind and relax in nature, but it can also be demanding physically. It is no small task to fill and move a wheelbarrow or bend over and weed an entire garden bed! At this point in the season, it is possible your skin is bug-bite covered and your muscles are sore. The vigor and excitement that pulled one from bed in the early mornings of spring may seem a distant memory. And many gardeners, not dissimilar from the flowers, are slowly getting ready for a day of rest, perhaps curled up in a soft blanket with a cup of tea.

Let us not forget there are so many more blooms to come; it's still summer! And the shift of color palette from later summer to autumn is something to behold. In an effort to conserve energy and find the balance between stillness and the work required of tending a garden, perhaps consider what I like to call slow gardening. This is the simple practice of being present with your body in a place of beauty and finding kindness for yourself in the process. Slow gardening can look like this:

» Noticing your breath
» Observing your landscape for a full year before making changes
» Touching the petals of a bloom before cutting it
» Focusing on a single bloom for an extended period of time; to begin, try five minutes
» Walking barefoot
» Arranging flowers and holding the person you are arranging them for in your mind's eye
» Pulling a weed on a purposeful exhalation
» Planning a sensory garden that focuses on smell, touch, sound
» Sowing the seeds of a flower that reminds you of someone you miss, oh so much
» Putting your hands in the soil, less garden gloves
» Admiring a bee asleep on a zinnia

Morning Meditation

Some mornings, I rise in an almost frantic fashion and stumble toward the garden. Cup of coffee in one hand, wrapped in a bathrobe, feet bare, utterly clueless of my earthly surroundings. At some juncture, between watering the dahlias and the perennial garden, I come to consciousness, seemingly confused by how I came to be outdoors. Squinting at the bright morning sun, I focus on whatever tragedy has befallen the garden while I slept—the aftermath of an excessive rain, or the beetles, or my inability to cut back the cosmos.

But when I am tired of this chaotic scene, I'm reminded of the importance of purposefully entering the garden—not for the purpose of productivity, but just for the lofty goal of observing the garden as it lives before me. If I can remember, I sit down before tending the earth and slowly feel the weight of the coffee cup in my hand, the warmth as I pull it close to my stomach. I recognize my breath and hold still for the soft dips of motion that roll through the flower stems tightly in rows.

I let my conscious mind return to my body and take long, purposeful sips of coffee. If you are like me, and the act of stillness is near to torture, be aware of this sensation. Can you stay just a minute longer than you thought you might? Often five minutes is about all I can manage before leaping from my seat and dashing for my watering can. But the mornings I pause for just a few short minutes, I'm rewarded with clarity of thought that lasts me the entire day.

Moon Walk

Certain flowers that are pale and light in color often go unnoticed in a garden packed with vibrant, brightly colored blooms. But at night, flowers like wild yarrow, Solomon's seal, and peonies absolutely glow. The structure of a garden reveals itself in harsh, abstract-like lines against the gray night sky. It is impossible to do tasks like weeding in the dark. Instead, you are left just to consider the grandness of the space and appreciate your hard work.

Stretch

Working in a garden can be incredibly difficult on the body, most notably the back, shoulders, and hips. Consider taking a mental scan of your body, searching for areas that feel sore and tired. Make time to gently move and warm up these areas by stretching both before and after gardening.

For the lower body, I make sure to do some deep lunges, supporting my lower back with my hands while paying close attention to the tightness of my hips. In gardening, we often sit and bend over for long periods of time, compressing our hip muscles. A lunge will help to open up this space.

To stretch my shoulders and open my chest, I use a rake as a prop, gently sliding it down my neck and placing it behind my shoulder blades. Next, I wrap my arms around the wooden handle and slowly twist side to side, raising my heart skyward and slowly breathing into my chest.

Finally, I warm up my back by lowering to all fours and transitioning between cat and cow pose. This is a simple yet highly effective way to not only warm up the body, but also ground into the soil before I take to the dirt.

Observe

Do you remember the dark, cold days of January when you created a list of flowers you hoped to grow and determined how many seeds needed to be planted? Perhaps it seems like a lifetime ago. I often have plans for a third succession of zinnias or cosmos that never come together. The empty rows haunt me a bit, but I know there is always next year.

At this time in the season, there is plenty to be done, and most of it seems impossible to finish. Harvesting, weeding, fertilizing, and a never-ending rotation of chores. Oftentimes it can feel like you are putting out one fire after another. Perhaps the idea of observing your garden seems like just another thing to add to the list. But I assure you this is very important.

Taking notes on the health and production of your garden is an invaluable practice. It will guide your design process in the coming years, and if you take notes now, it will be much more accurate than in November. If you have been walking your garden daily, most of this information will quickly come to you. If you haven't, let this be an opportunity to return to the soil and slow down.

If you can, try to remind yourself that this is not another chore but something to be curious about. Perhaps go out as the sun is setting, in the golden hour, or first thing in the morning when all is quiet. Remember those colored pencils we used at the beginning of the year to draw blooms? High summer is a wonderful time to sketch your flowers as they are in the moment.

High-Season Observations

- What pests have you noticed this season, and which blooms were most impacted?
- What have been some effective pest-management strategies?
- What flower did not perform as well as you had hoped? Any thoughts on why?
- What flower do you wish you had planted more of?
- How many blooms per week do your cut-and-come-again flowers produce on average?
- What are some blooms with incredible vase life?
- Blooms with a disappointing vase life?
- Blooms you have observed in other people's gardens or photos that you would like to grow next year?
- What elements of growing flowers felt overwhelming to you?
- What floral successes did you notice in your garden?
- Thoughts on expanding your growing space next year or perhaps scaling it back?
- What seeds do you have left? (This is for those of you who, like me, forget and order far too many seeds.)
- Favorite flower combinations in bouquets?
- Was there any time during the season that the garden felt flat or a lack of blooms was noticeable?
- How has your irrigation system fared?
- Are there systems you could put in place to work smarter, not harder?

Autumn
Closing & Reflection

The garden of autumn is gold and evergreen, full of seemingly wise flowers that stand upright, twisting toward the morning sun until the very first hard frost. Not as youthful and screaming yellow as the daffodil or as showy as a peony, not quite as wild as meadow rue, just deeply still and bold, with nothing to lose. As the temperature drops, flower stems weaken. You might walk out to the garden and notice a 'Café au Lait' dahlia dangling upside down, its stem unable to hold the luscious petals any longer. Birdsong fades gradually, but likely you don't consider this until one morning it is the absence of sound that strikes you. Finally, it is possible to again move and shovel the soil without sweat dripping down your back. This means the more physically taxing jobs, covering beds with mulch or natural organic material, cutting down spent blooms to the earth, and weeding, are somewhat manageable again.

Autumnal Rhythms

Autumn is a time of reflection. The first killing frost reveals the bare outline of your flower stems, the bones of a garden. With little visual distraction, consider what might be possible in a given space the following season. Any new garden bed preparations completed at this time of year reduce the amount of work required before planting precious seedlings in early spring. It is also a time to be honest about where to reduce your energy. Without judgment, consider substituting a row of annuals for low-maintenance perennials. As the weather is cool and damp in the morning, most of my garden chores wait until the air warms in the afternoon. But time is short, and the sun starts to dip below the tree line long before I'm ready to go in for dinner.

The noticeable shift to the closing chapter of your garden can last anywhere from four to six weeks, depending on the first killing frost. This is a large window of time you might maximize by adding perennials that produce lovely late-season cut flowers or creating bouquets and arrangements with the swan song flowers of autumn. Things like Solomon's seal and raspberry branches begin to turn yellow and orange, making them appear a new plant entirely. Native blooms, like aster, pop up on the roadsides and in untended areas of the yard. Keep your eyes open and alive to the changing landscape in front of you. After months of weeding and watering, we often forget to enter our garden with new eyes.

As the length of the day shortens, you will notice the flowers take longer to come into bloom or put on new stems. There is a sadness in watching a garden slowly fold in on itself and fade into the earth. I usually describe the closing of a garden, the weeks leading up to the final frost, as "tucking things into sleep," and wander the yard creating a mental plan for the weeks to come. It is also a time to focus on soil health and add suggested amendments before everything is under snow.

If you are tired after high summer, don't let the idea of putting a garden to bed feel insurmountable. In truth, you could always just close your eyes and hope for the best. Plants are incredibly resilient, and weather can be unpredictable. But a little extra time spent tending the garden in the fall means you will be off to a good start come early spring.

How to Close Your Garden

Set aside a crisp sun-filled morning to practice the art of observation. Enter the garden with a notebook and pencil, and begin a slow, purposeful walk. Start by calling attention to your breath and the sensation of your feet anchoring in the soil. Bend to the earth and observe each flower with purpose. Do you notice yellowed leaves or signs of pests? Jot down how the soil feels. Is it soaking wet or dry? Depending on how they fared over the summer, certain garden beds may need more organic matter or mulch. Do any of your blooms look sickly? Plants that appear to have a fungus or look generally unwell should be ripped out and disposed of separately from healthy organic matter that ends up in the compost. Examining your garden closely will help you recall details from high summer you may have forgotten. If there is a flower you've decided isn't a desirable bloom for next season, record the reason why and the location of this crop in the garden. In recent years, I have given up on ranunculus completely and made sure to note precisely why I made this decision. This way, when the gorgeous catalogs appear in late winter, and I'm gushing over this buttercup family beauty, I have the willpower to move on and consider something else.

WEEDING IN THE FALL

At this point in the season, weeds, just like flowers, will go to seed and start self-sowing. Removing as many weeds as possible in autumn will prevent them from popping up next year. Also, if you have fallen behind on weeding, as I often do, take a close look at the perennials. If these blooms are wrapped in weeds, it is incredibly difficult for each plant to maintain a healthy root system. Therefore, focus on weeding your perennial flowers first to protect these precious long-lasting garden favorites. An annual bed full of weeds can be easily turned over in spring using a broad fork and moving soil with a shovel, so push this farther down your to-do list.

CUTTING BACK THE GARDEN: YAY OR NAY?

People often inquire about cutting back their gardens at this time of year. In truth, not all flowers need to be cut back. I keep the majority of my blooms intact, allowing the branches to darken and develop into sharp lines that crisscross the sky in an artful fashion during the long nights of winter. Leaving the remains of your flower garden in place is also beneficial for wildlife. In our front garden, it is not uncommon to see a chickadee in a snowstorm feeding off a spent sunflower. Watching this event with a cup of tea is truly heartwarming. However, if you notice any flowers with signs of mildew (irises, peonies, and daylilies are particularly prone to this), absolutely cut them back and remove the cuttings from your garden or compost pile. Keep your eyes open for any spent plants prone to developing a mushy texture in the colder months. If the leaves or stem are easy to squish, cut them back and toss them out.

THE POWER OF MULCH

Organic material and mulch are what I consider to be the metaphorical blanket under which to tuck your precious flower roots for a long winter's rest. Consider raking all the leaves you can muster and placing them on your garden beds, or use organic matter such as compost, wood chips, and/or straw. If the soil starts to look tired by season's end, you might add compost or loam. Properly insulating your flowers protects them from frost heaves in the harsh winter.

AUTUMNAL PRUNING

Evaluate which shrubs need to be pruned at this time of year. The list should be small, as most should be cut back directly after they flower. However, some hydrangeas, such as *Hydrangea arborescens* 'Annabelle' and *H. paniculata* 'Limelight', can be pruned in the fall. Lavender also benefits from being cut back roughly by one-third before the first frost. Consider harvesting lavender and hydrangea at this time for drying purposes as well.

Taking Stock of the Garden

After recording your observations, reflect upon what you have learned by focusing less on individual blooms and more on the garden as a whole. Consider moving certain crops to different locations, decreasing the likelihood of pest damage in the years to come. If you want to add more flowers, review any spaces that have opened up in your garden beds after adjusting your grow list for the following season. Are there areas where you could work in some interplanting or bits of lawn that could be transitioned into garden beds? Of course, you will have all winter to scheme and dream, but reviewing plans while physically in the garden with your current flowers is helpful.

Late-Season Blooms

One of the most difficult things about planning a garden is ensuring there will be color and interest all season long. Often the late-season blooms are forgotten or an afterthought. If you have the time to visit a proper botanical garden or the garden of an experienced gardener, you will likely notice the garden appears full and lush up to the final frost. Do be sure to jot down what varieties hold your interest, and perhaps slowly invest in perennials that bloom into late fall. This is just a practical suggestion, as often the last thing on your mind in July is sowing new seedlings indoors to transplant for a harvest in September. However, a perennial that continues to bloom straight through until the first frost requires virtually no effort. Here I'll share my favorite late-season blooms.

CHINA ASTER (*CALLISTEPHUS CHINENSIS*)

This flower comes in a variety of colors, from deep dark purple to light pink and white. Unlike other asters, China aster is an annual and prefers moist soil with full sun to part shade. It is excellent to add structured texture to a bouquet and easy to match with a desired color palette as it comes in so many different shades.

DAHLIA

This bloom is the current darling of the cut flower world, so much so that the inspiring flower farmer Erin Benzakein published an entire book on it, entitled *Floret Farm's Discovering Dahlias*. Here in New England, it often seems dahlias have just hit their stride when a killing frost comes in and gobbles them all up. However,

these flowers produce so many stems that even a few weeks with them in your garden is incredible. Invest in three or four varieties to start, making sure you find a collection that is entirely different in shape but has colors that complement one another. Most dahlia tuber sales begin in early winter, and the more popular varieties tend to sell out. I recommend finding a local dahlia flower farmer and adding your name to their mailing list. Supporting local agriculture is something I feel strongly about, and I have found flower farmers to be generous in sharing helpful growing tips.

JAPANESE ANEMONE (*ANEMONE* SPP.)

This bloom has delicately petaled white and pink flowers and can grow tall, from three to five feet, so consider putting them in the back of a garden. Keep well contained as it can spread rapidly. A perfect bloom for a spot in your yard with moist soil; it's also deer-resistant.

MARIGOLD (*TAGETES* SPP.)

I grow lots of giant orange marigolds in the garden. While they do bloom all season long, to me they stand out at season's end, as their color is cheerful and bright right up to the very last frost. At this time of year, I often harvest marigolds to create natural dyes and have found they reliably put on blooms after being cut late in the season. A word of caution: bees seem to love marigolds! As such, I have been stung many times when harvesting in a hurry. Let marigolds and their pollinators remind you to slow down by checking for bees each time you lay your hand upon a bloom. Plant in full sun, in dry to moist soil.

A Word on Season Extension

With the arrival of cool nights and shorter days, how we might hold on to our cut flowers for even just a few days longer becomes a running line of thought. Many blooms will tolerate a few light frosts and continue to produce blooms, but it is the killing frost that truly destroys a garden. Occurring at roughly 24 degrees and below, this frost blackens stems and destroys the root system of annual flowers. It is difficult to predict exactly when this will occur in your garden. Of course, the last frost dates we identified earlier in the year should be a good guide. So how does one potentially extend the harvest of their cut flower garden?

Here at the farmhouse, I dust off my garden hoops and put them in the ground before the soil becomes too hard from cold. A killing frost is difficult to avoid, but freezes prior to this, between 25 and 32 degrees, can be thwarted. Using metal garden hoops and row covers can extend your season for a few weeks but be sure to cover at night and then pull off the cover during the day to provide as much sunlight as possible.

Season extension is also important in the spring. I use hoops and row covers for cold-hardy flowers such as larkspur and sweet peas, which are planted out earlier than other blooms. If you take a risk and plant these flowers too early, you can cover them when the last frost or two rolls through. I also use a light frost cover to protect my tulips from deer, who are notorious for eating these flowers just before they open up.

For super beginners, you may want to wait on extending the season for a couple years while you get the hang of the garden's rhythm. That's okay; remember that your pace is your pace, and rest is a beautiful thing.

Tools for Season Extension

It can be difficult to say goodbye to a garden, and every last flower plucked from the earth seems a small miracle. If you'd like to try and add a few weeks to the end of the season and enjoy your blooms for just a bit longer, consider the methods listed below. Most of the items are affordable and will last for years. I suggest investing in the wire hoops and frost cloth before bothering with heavier duty methods of frost protection. Often these supplies are offered in quantities too large for a home gardener, so consider sharing them with a neighbor.

SIMPLE WIRE HOOPS

Wire support hoops are a great starting point for season extension. They are roughly a foot and a half high and two feet wide. Make sure to space them five to eight feet apart in a row. I use wire hoops frequently for not only season extension but to drape shade cloth or frost cloth to use as pest protection. These hoops are easy to store and carry, but do not hold heavier fabric like Agribon (see below) effectively.

HALF-INCH GALVANIZED ELECTRICAL CONDUIT AND HOOP BENDER
An affordable way to create heavy-duty hoops is by purchasing galvanized electrical conduit from a hardware store and a low tunnel hoop bender from a farm supply store. The trick is to stabilize the hoop bender on a table or wall (or by having another person hold it), slide the conduit through the bender, and pull. At first, it can feel awkward, but once you get in a rhythm, you'll be surprised how easy it is. This is a very affordable way to create a low tunnel that will hold the entire winter or simply to use them as hoops that will handle heavier frost fabric.

LIGHT FROST COVER
A light frost cover can be purchased from a local greenhouse or an online garden supply shop, but you could also use a sheet or lightweight blanket for the job. This works well when the temperature drops to around 32 degrees, and they are easy to toss on and off. Wire hoops will hold this type of frost cover without issue, but in the evening, if there is wind, consider holding down the edges with anything you can find in the yard, like fallen branches, old bricks, or rocks.

AGRIBON
This frost cover is offered in various strengths to thwart crop damage at different temperature levels. The higher the number, the more frost protection it affords, but keep in mind less sunlight will get through to your blooms. I use Agribon 70 for blooms I have overwintered, and it provides protection down to 8 degrees. Since I have this fabric on hand, I also use it when a hard frost or freeze comes through. However, for most home gardeners, a lesser-strength Agribon is perfect for simple season extension.

It's Time for the Dahlia Dig

Many beginner gardeners I have worked with admit that the idea of digging up a row of dahlias on a cold day in October hangs over their head. That is until the ground freezes and they experience a wave of guilt that their precious dahlia tubers will now rot! Go easy on yourself, and start where you are. Just dig up one single dahlia. Or walk to the garden with the intention of digging one single dahlia and be open to the idea of digging more if your body and mind allow.

Perhaps the best time to dig dahlias is when you feel a bit frustrated, tired, or stuck. I have always found that moving my body often results in a shift of mind. Start by cutting the entire dahlia stem, leaving roughly six inches sticking out of the soil. Use this bit of stem like a handle, wiggle it back and forth, and gently lift your plant out of the earth. Pause, take a deep breath, and observe the magic that has occurred under the soil all summer long without your noticing. Let this remind you that good things happen even when you forget to take the time to slow down and see them.

If you dig more than one dahlia, you'll be amazed how quickly you can accumulate dozens of new tubers for the following season. One dahlia can often turn into eight or ten in a single year, allowing you to grow buckets of blooms for years to come or share your collection with others.

But remember, you can simply start where you are, save that big clump of tubers from one single plant, and continue on with your day. Also, it is not necessary to dig, clean, and store all on the same day. Personally, I find digging dahlias fascinating. While beautiful blooms were churning out for weeks above the soil, beneath the soil tubers were multiplying. It can be shocking to put your shovel in the soil and lift out a dozen tubers created over the growing season.

How to Dig Dahlias

» Dig your dahlias after the first killing frost or when the plant begins to brown and die back.

» Cut back the dahlia until there is about four to six inches of plant remaining.

» Use a spade to dig gently in a circle approximately eight inches away from the plant. If you dig too close, you might puncture a tuber, causing it to rot.

» Use a pitchfork to loosen the soil beneath the plant. Then, gently rock the remaining stem of the dahlia back and forth, pulling it slowly from the soil.

» Inspect dahlias for signs of rot or crown gall. What does crown gall look like? This bacteria causes an infection that produces bulbous white growths. If you find crown gall, throw away the tuber, taking care to dispose of it far away from your gardens and compost. Next, wash your tools in soap and hot water before continuing your dahlia dig.

» Should any tubers become cut or damaged from the dig, toss them out.

If you are waiting to divide and store your dahlias, don't rinse off the dirt and place them in a large paper grocery bag. Store this bag in a dark, cool space until you are able to clean and divide.

Dividing and Storing Dahlias

The most complicated bit of storing dahlias is finding the proper location. This space needs to stay above freezing without getting too warm, maintaining a temperature of roughly 35 to 50 degrees. A garage, basement, or attic usually fits the bill. For people who don't have an area that stays in the desired temperature range, consider storing dahlias in a spare refrigerator.

Remove all dirt by using the pressure-washer attachment on your hose. Lay the tubers out on the ground and spray them down until most of the dirt is removed. Be careful not to make the water stream too concentrated as this can damage tubers.

Let the tubers dry out on a sheet or paper bags indoors, out of direct sunlight. This usually takes a day or two but be sure the tubers aren't so dry they begin to shrivel. Essentially you just want them to be dry to the touch.

Once dried, use a paintbrush to gently flick off any dried dirt still on the tubers.

Take your snips and remove any tubers from the clump that have cracks or dents in their outer skin. Toss these into the compost.

Begin to remove individual tubers by locating those with a viable eye. This small raised white dot at the narrow end of the tuber is where your new dahlia will emerge. Locating the eye can take some practice, so make sure you are working somewhere with good lighting.

Cut individual tubers off, working carefully not to damage other tubers and ensuring that the eye stays intact.

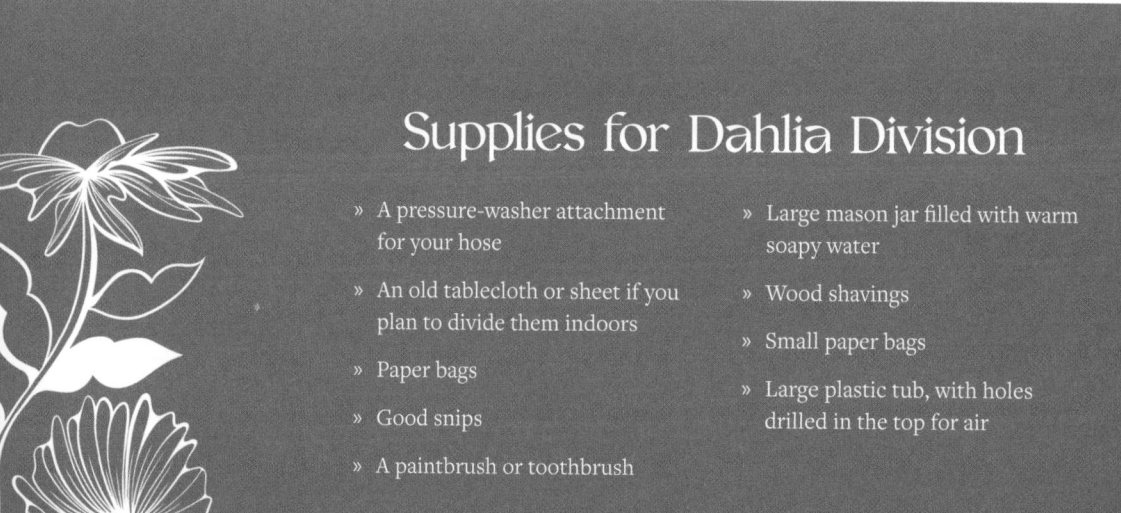

Supplies for Dahlia Division

- » A pressure-washer attachment for your hose
- » An old tablecloth or sheet if you plan to divide them indoors
- » Paper bags
- » Good snips
- » A paintbrush or toothbrush
- » Large mason jar filled with warm soapy water
- » Wood shavings
- » Small paper bags
- » Large plastic tub, with holes drilled in the top for air

It is nearly impossible to perfectly remove every single tuber from a large dahlia clump! Inevitably, removing one will damage another. So focus on cutting and saving the tubers with the most obvious eyes. Remove any little bits of growth that look like small dirty whiskers; these thin pieces are prone to rot.

Once the tubers have completely dried, which may take a day or two, place them in small paper bags with wood chips. I individually bag each tuber; this way, if one rots, it does not destroy the others.

I place these bags in a large storage tub, and make sure to drill some air holes into the lid.

Check your tubers once a month by slowly unwrapping each bag and looking for signs of rot. Throw away any tubers that have mold or feel soft to the touch.

Cleaning Your Tools

After the killing frost, when there is no longer an outdoor garden to tend, one can turn toward tasks that were left woefully neglected during high summer. Perhaps the most important one is tidying and organizing all your garden supplies. Harvest buckets will need to be washed and stacked, drip line and frost cloth rolled and tucked away, and the tools! The tools all need a deep cleaning and oiling.

Good-quality garden tools can be expensive, but if you are diligent about cleaning and storing them properly, they are a wise investment. I'll be the first to admit I'm not the most reliable

gardener when it comes to cleaning gardening tools during the high season. My small form of apology is a good, honest clean once the season is over.

Hint: a tool-cleaning day is always more pleasant when it's sunny, so, if possible, check your weather report and do this at peak temperature.

Creating a New Garden Bed for Spring

If you have decided to expand your growing area in late fall before the snow arrives, consider solarization or occultation (also known as tarping). Both of these methods are easier on your body than digging. They also can be used to control weeds in your current annual beds at any point in the season.

Solarization requires heating the temperature of the soil to kill weeds and pathogens. To do this, place a sheet of clear plastic on top of your turf or overgrown garden bed, allowing sunlight to pass through, thus warming the soil. In dry, sunny places, solarization can kill vegetation in two to three weeks during the summer months. However, if your climate is cold or there is a long stretch of rain, the process can take longer.

How to Clean Your Garden Tools

» Create a cleaning station by setting up a worktable near a spigot, and consider using a pressure-washer attachment with your hose. This attachment allows you to adjust the pressure of your water stream and release dirt that might be packed onto tools like a broad fork.

» Round up all of your buckets and rinse them with natural soap and water. While the buckets are drying, clean any larger tools by spraying them down with water. Allow them to dry completely in the sun or with a towel before packing them away for winter.

» Collect your handheld tools and place them in a crate or on the grass, reduce some of the pressure on the hose, and give them a good spray, washing away any dirt.

» After the tools and buckets have been thoroughly rinsed, take one bucket and fill it with soap and warm water. Place your handheld garden tools in this bucket and let them soak for about an hour. Instead of letting the tools air-dry, leaving room for rust, wipe them down with a cloth.

» While drying each tool, closely examine the hinges, which are prone to rust, assess their sharpness, and look for sticky residue that did not come off in the cleaning process. When each tool appears sufficiently clean, take an old cloth and gently massage camellia oil into the blades and hinges. For tools with a wooden handle, check for any rough spots and sand them down until they are smooth. Finally, rub the handle with oil. If there are tools that feel especially dull, I will sharpen them with a file.

» Once the tools are cleaned, oiled, sharpened, and dried, wrap them in a soft cloth (an old shirt or sheet will do) and store them in a dry place for winter. A waterproof box tucked up on a shelf in your garage or shed should do the trick.

Occultation is the method I prefer for my garden. Instead of using a clear tarp, I use dark tarps or cardboard. The dark tarps absorb light but do not pass the heat down through the soil, thus the process takes longer, roughly four to six weeks. I find occultation is often easier as you can reuse tarps and cardboard you already have on hand. Instead of folding and tucking those tarps away for winter, put them out in the garden.

Occultation in Fall

To create a new garden bed, first measure and stake the area you would like to grow on. Record the measurement of the bed so you have these details on hand during the winter months while plotting your flower placement. Next, mow down all the vegetation in your new garden bed area and place a tarp over this space. Secure the tarp by weighing it down with heavy items from around the yard. In the past I've used logs, boards, branches, and cinder blocks. Leave the tarp in place for at least four to six weeks or remove once the snow has melted in early spring.

Note: this does not look attractive! Your lawn may have a bit of a junkyard feel, so if visual aesthetics are important and your garden is in the front yard, this may not be your method of choice. However, I have found that all is forgiven if by high summer you have beautiful rows of blooms to share with your neighbors.

Using occultation is an easy way to kill weeds in an annual bed. Putting this process into place early in the spring, as soon as the soil can be worked, will kill the first generation of weeds awaiting you beneath the soil. The process in spring is exactly the same as in fall, with one important exception. Just before you place the tarp on top of your garden bed, heavily water the area, ensuring that the germination of the weed seed occurs. After this initial watering, secure the edges of the tarp to trap in heat and moisture. Approximately four to six weeks later, aerate the garden bed by using a broad fork, work in some fresh compost, and plant your new seedlings in the soil.

Working with Dried Flowers

My hands often feel idle while the garden sleeps. I alternate between gesturing wildly to tell stories and holding them, fingers clasped quietly, in my lap. One's fingernails can seem painfully sterile when not a bit of earth is beneath them. And there is a slowness when you wake in the morning. No longer are mornings spent tumbling out the front door while pulling on boots to water some delicate stemmed blooms. At this time of year, I rise slowly and stare at the moon, often searching for Venus with my naked eye.

The abrupt end to the gardening season can leave me with a touch of melancholy, so to stave off winter blues, I've cultivated flowers that might be used indoors all season long. Creating with dried flowers, or foraging for natural elements in winter, has given me a year-round connection to the garden. I would never claim to be a crafty person and in truth many things I create are more about the process and not a finished product. For me, the joy is in the experience of holding a dried bit of strawflower in my hand, delicately pulling the petals, and placing them on a blank sheet of paper. Much like the way I function in a high-summer garden, I move slowly, calling attention to the faded colors of each bloom in front of me. This practice reminds me to appreciate the small beauties in life that can often go unnoticed.

Pressed Flowers

Pressed flowers often fall into two categories: herbarium-inspired and artful. The more traditional herbarium-inspired pressed flowers, complete with important details such as the location of species collection and the proper Latin name, are commonly used in academia or research settings. I often press flowers in this fashion, not for an academic purpose, but simply because it looks beautiful.

If you have a moment, do find your way to the digital images of the Emily Dickinson herbarium, which are accessible through the Houghton Library of Harvard University. While alive, Dickinson was more known for her gardens than her poetry. In her journal, you will notice each specimen placed just so, with the utmost care and attention, akin to her use of language in her poems. It is easy to get lost viewing hundreds of her pressed blooms. One can almost imagine walking alongside her as she kneels down to harvest an iris. If this type of design inspires you, consider purchasing a herbarium stamp or print a few herbarium label templates. I like pressing single flowers from special places, or moments in time, in this style. It is as if you've captured a bit of sentimental beauty.

Creating artful designs from pressed flowers began in Japan, a practice known as *oshibana*. By working slowly, the artist is encouraged not only to create something beautiful but to hone their patience. Dried flowers are so very delicate that I find when working with them the world naturally slows down and one can become hyperfocused on the smallest of petals. If you struggle with patience, consider giving yourself time to explore in this way.

Pressed flowers have seen a resurgence in popularity in recent years. In the wedding world, brides often look for their bouquet to be pressed as a keepsake. More recently, fabric has been designed with images of pressed flowers and jewelry with small pressed flowers can be found in many boutiques. Working with pressed flowers is quite simple. While there are large scientific-grade flower presses, I have always relied on the age-old method of pressing flowers in a book. There is a bit of romance in this process.

How to Create with Pressed Flowers

» Harvest flowers that are known to press well. As a general rule, blooms with a single layer of petals work best. Consider cosmos, violets, hydrangea, astrantia, Queen Anne's lace, ferns, nasturtium, and aster.

» Harvest your flowers after the morning dew has dried off the petals. Never cut flowers for drying if they are wet from rain, as this will lead to brown spots visible after the drying process.

» Line the pages of a large, heavy book with wax paper or newspaper. Gently place your flower on top of the wax paper and slowly consider the arch or angle of each bloom. Once you have found just the right positioning, slowly close the book. Place an additional book or two on top for good measure.

» For a large collection of flowers, consider taking pieces of cardboard, lining them with wax paper or newspaper, placing your flowers on top, and weighing the top layer of cardboard down with cinder blocks.

» Every few days, open the pages of the book and gently touch the petals of the flower, looking for any signs of moisture.

» Once flowers have completely dried, wrap them in tissue paper and store them out of direct sunlight and away from moisture.

When working with dried flowers, I find handling them with tweezers gives me more control, as they tend to be finicky. Often, I'll lay out all the flowers in their desired positions and take a photo with my cell phone. This way if I sneeze or bump something with my elbow I remember where everything belongs. Determining where to place a single flower or petal takes time. No need to rush the process.

Using a Q-tip or small brush, gently glue each bloom to the piece of paper you have selected. Any craft glue you have on hand will do; be sure the glue has dried before pressing it into a frame.

Consider adding pressed flowers to a letter or card and send it to a friend you miss dearly. A small pressed flower can offer a lovely personal touch, especially when it's from your garden. While traveling, I've been known to pick a few flowers that catch my eye, press them in a book during our journey, and add them to a herbarium journal when we return home.

Finally, I like to use pressed flowers to create a small reminder of the garden each year. While there have been plenty of seasons I've forgotten, one of my most prized possessions is a small, unassuming frame that hangs near the kitchen. Pressed inside is a handful of flowers from the very first garden my husband and I shared over fifteen years ago. If this is your first year growing a garden, do set aside some flowers to create a special keepsake! You will be amazed to see how your collection of blooms increases over the years.

Cyanotypes

Using pressed or fresh flowers to create sunprints, more formally known as cyanotypes, is a simple way to capture the fleeting essence and movement of a flower forever. Perhaps the most well-known cyanotypes are those of algae created by botanist and photographer Anna Atkins. Her print of *Dictyota dichotoma* from her book *British Algae: Cyanotype Impressions*, self-published in 1843, is so bright white that you can hardly imagine this bit of seaweed is brown in the natural world.

The cyanotype process involves treating paper or fabric with a combination of potassium ferricyanide and ferric ammonium citrate, arranging the items you wish to copy on top of the paper, and exposing all items to ultraviolet rays. While this sounds complicated, fear not! One can purchase pretreated paper and fabrics if mixing the chemicals at home feels too much. When your paper arrives, pull it from its light-proof black bag, lay your pressed flower on it in an artful way, and expose the paper to bright sunlight in a protected location with little wind.

After about five minutes (the time can vary based on the strength of the sunlight), wash the paper in cool water and let it air-dry. You will be left with the deepest blue background and a stark white outline of your pressed flower. I recommend trialing this process with delicate blooms like Queen Anne's lace, ferns, and poppies.

Everlasting Flowers

The area designated for dried flowers in my cutting garden has increased in size year after year. This is partly because I am more dedicated to purchasing only local flowers, and if I want local flowers in the winter months, the only option is dried. But just as important, I adore the muted tones that dried flowers create and the gentleness you must apply when designing with them.

    The use of dried flowers has been around since we were able to cultivate seeds. Dried flowers were used in the tombs of Egyptians, to decorate champions in Greek and Roman times, and to cover offending smells before modern plumbing arrived. In the 1960s, dried flowers began to fall out of fashion. This was for two reasons: first, the introduction of plastic flowers, and second, the globalization of the flower market. Suddenly, if you wanted roses in February, you could have them.

Recently, dried flowers have had a bit of a renaissance and even been rebranded as "everlasting flowers." It is not uncommon to see weddings adorned with dried florals or retail shops that decorate their window displays or exterior entrances with dried elements. Wearable dried flowers have also returned as necklaces, hairpieces, and pins.

Flowers That Are Good for Drying

For most flowers, the drying process is simple. Harvest your blooms, tie them together with a bit of twine, and hang them upside down in a dark closet or hallway with good circulation for roughly two weeks. After this, you can wrap the flowers in tissue paper and put them in a container to keep them safe until you are ready to use them. Not all flowers work well as dried flowers; below is a list of some of the blooms I've worked with over the years. All of these varieties can be dried following the method above.

ASTRANTIA

This star-shaped perennial beauty fades in color once dried, but the distinct shape holds so well that it adds interest to any dried creation. Harvest when the stem is stiff to avoid drooping.

COCKSCOMB

A textural flower that adds just the right amount of color to a dried arrangement full of neutral elements like grasses and seed pods. Cockscomb is a prolific grower and very low maintenance. Harvest when all the flowers have opened but before they brown on the stem.

'FROSTED EXPLOSION' GRASS

This grass has a glittery plume that often seems overly wild in a high-summer bouquet or arrangement. Harvest when the grass has just barely opened and is still slightly green; it will hold better this way during the drying process. Harvesting after the grass has blown open means it is less likely to stay in place when you design with it months later.

GLOBE AMARANTH

My intention when first planting globe amaranth was to use it as a fresh cut flower but I found, no matter what color I grew, it failed to create a harmonious color palette with other flowers in a bouquet. However, after drying these blooms, the muted tones elevated the appearance of dried grasses or seed pods that seemed one note. Harvest this bloom when all the flower petals have unfurled.

NIGELLA

It is hard for me to decide whether to cut this bloom as a fresh flower or to save it for drying; both options are beautiful. Harvest after all the petals have dropped off the bloom and the seed pod feels hard.

PINCUSHION FLOWER

This bloom functions as a perennial or annual and has tiny, delicate flowers. After the petals drop, harvest and save for its cheerful seed pod.

STRAWFLOWER

This flower looks like a dried flower even when well hydrated and standing tall in the field in peak summer. Harvest when the exterior petals open in the morning sunlight, revealing their colored seed center.

YARROW

A hardy bloom that holds color well over time. I often create dried flower posies with yarrow as a base layer because it isn't prone to snapping. Harvest when all the petals are near to open; waiting too long results in browned flowers.

Creating with Dried Flowers

Once you have grown, harvested, and dried your flowers, it is time to create with them. Many of the holiday gifts I give to friends and family are made from dried flowers grown in my back garden. When you work with dried flowers, remember to press lightly and breathe deeply. It always seems that when I have placed a dried flower just so, the stem snaps. When this happens, simply take a deep breath and begin again. Consider this activity an excellent way to slow down and focus on a single task. Below is a list of some simple crafts I've created at home.

POSY

I just adore a small posy tied together with four to five dried elements. I find it is a sweet small gesture and an economical way to share bits of your garden year-round. In Victorian times, many well-to-do women had what was called a tussie-mussie.

This small vase would hold your posy and connect to your hand with a chain, allowing one to dance uninhibited while displaying their blooms. While the tussie-mussie has long fallen out of fashion, I enjoy sending a small posy of dried flowers to friends by attaching them to a bit of cardstock with decorative washi tape. Making a posy is quite simple: take a small collection of dried flowers (often those blooms whose stems are too short for a wreath) and wrap them together, tight, with floral twine. Determining how tight is too tight can be difficult and you may lose a few flowers before finding the ideal resistance. Strawflowers have unexplainably flimsy stems and it seems as if I'm always popping the flower head off when working with them. Should this happen to you, cut a bit of floral twine, poke it through the center of the bloom, loop the end of the twine, and, once it has emerged from the flower, pull it back through the seed head slowly, with control. This creates a strong faux stem of sorts. Eventually you will administer just the right amount of pressure to all your dried flowers—it just takes a bit of practice!

BELL JAR ARRANGEMENT

Using a bell jar with a wooden bottom, anchor a floral frog with glue or clay onto the base. Arrange dried flowers as you would fresh, keeping in mind these blooms will stick straight up and offer very little in the way of arch or bend. In my opinion, less is more in an arrangement like this. Be sure to leave room for each bloom to tell its own story.

HAIRCOMB

Using a bit of floral twine and approximately a dozen dried flowers, one can create a unique haircomb. Make two to three very small posies by wrapping three or four flowers together, then secure each individual posy to the comb. To do this, simply weave another piece of floral twine in between the teeth of the comb and over the posy.

SCULPTURAL PIECE

A few years ago, I had an excessive amount of ornamental grass and wild aster hanging upside down in the back of a closet. With a bit of time on my hands, I decided to form a ball of chicken wire and hang it from the ceiling. As the afternoon sun dipped low in the sky, I gently placed bits of dried elements into wire, resulting in an organic disco ball of sorts. The trick is, don't think! Turn your mind off and practice nonjudgment. The wilder, the better with this type of design. Add to the hanging ball over the course of the winter by pulling little elements from the woods or roadside while you walk. Take it down from the ceiling and use it as a centerpiece when friends come for dinner.

STRAWFLOWER GARLAND

A simple cheerful garland is an eye-catching organic touch on a bookshelf or above a bed. Strawflowers or dried daffodils are my go-to for creating an everlasting floral garland. Take a long bit of twine and a collection of strawflower blooms, less their stems, and thread the twine through the center of the flower, repeating this step over and over again until you reach your desired length. I find it can be challenging to press twine through the hard seed center, so often I poke the center with a small screwdriver. When you have finished stringing your garland, add some dyed ribbon for a bit of revelry.

Embracing and Creating Tradition

In November and early December, when the nights can feel oh so long, I find myself pulling on a long sweater and moving slowly. The garden is an unrecognizable heap of browned sticks and blackened sunflower seed heads poking up at random intervals throughout the backyard. I watch from the kitchen window as birds gracefully move between zinnias and nibble on the very last seeds of autumn. It is during this time period that I wander the woods or the seashore, foraging for bits of greenery and other natural items to create seasonal decorations that speak of place. Being in touch with natural elements that remind us of the world beyond our frosty

window is essential for our mental health. Just smelling or touching bits of foraged greenery can lift one's spirits.

Bringing greenery indoors during the winter months is hardly a new tradition. During the winter solstice, ancient Celts would light fires and bring bits of pine, evergreen, and holly into their homes to ward off evil spirits. They believed evergreen possessed a certain magic as it stayed alive and vibrant long after other plants perished during the first hard frost. In Victorian times, this tradition evolved and people decorated their mantelpieces with holly, herbs, ivy, berries, and mistletoe. Leftover tree clippings were repurposed for home decorations at a time when nothing was wasted. Essentially, anything that was green or smelled lovely was gathered up and placed in a single room, most often the one with a fire.

While my wreaths tend to be less about eternal life and more about grounding in my present environment, I do believe finding meaning or purpose while creating with natural elements is important. During this time of year, we often become hurried, anxious, and buried by the long list of things to do. The stillness of harvesting and creating with elements acquired by taking a long, slow walk in the woods

Natural Items for Winter Design

- Balsam fir
- Cedar (Atlantic white and northern white)
- Eastern white pine
- White spruce
- Boxwood
- Hydrangea
- Sumac
- Dried aster
- Dried flowers
- Milkweed
- Roadside weeds
- Seashells
- Seaweed

is the sort of tradition I appreciate. Bringing organic elements into your home that speak of place is less about beauty and symmetry and more about letting your mind wander. Extra points if you allow yourself to create yours in dim lighting, as the sun sets, perhaps by the glow of a candle.

A few final notes about sourcing products for your winter decorations. It's always good to be mindful and avoid pulling red berries from invasive plants (bittersweet, for example). Before cutting branches off a tree, look for elements that may have already fallen on the forest floor. I often search for things like pinecones and snapped branches after a big windstorm.

Remember to consider sustainable practices and perhaps repurpose old fabric to be used as ribbon, or tuck bits of dried seaweed into your design here and there. Not only is sourcing locally better for the environment, it is more forgiving on your wallet. My humble opinion is that we've gone a bit overboard with holiday decorations, myself included. There tends to be less emphasis on bringing the natural world in and more focus on buying armloads of holiday decor from a big-box store.

I'm not suggesting you toss all your precious memory-fueled holiday items. Instead, perhaps see if you can weave natural elements into your home bit by bit each year. I encourage you to also be mindful of the impact natural living greenery has on your well-being, versus the plastic alternative.

Here I offer a list of some of the items I source when creating designs for the home during the winter season. With most of these items I find less is more—allowing a single stick or branch to tell a story is powerful. Also consider that plenty of these items, such as sumac, are beneficial to wildlife, so be sure to take small amounts and only what you truly will use.

Making a Seasonal Wreath

I've made lots of different wreaths over the years, but my favorite is a simple modern metal ring dressed in natural elements, secured with twine. I like that a wreath in this style doesn't scream "holidays" and can be hung on the wall all winter long. Often, I sneak little pieces of interesting natural items into the wreath for months. Reminding myself to look for these little organic treasures helps to keep my mind focused on the natural world in front of me. Thus, it is a constantly evolving bit of whimsy during the dark days of winter.

    To make this type of wreath, take a slow walk in your backyard or in a favorite park. Look for any organic elements that may have blown onto the ground and into your garden over the past few months. Consider the curve and shape of bare branches and tread lightly, picking up each acorn or admiring the snapped stem of a browned wild aster blowing in the breeze. Return inside and search for any dried flowers you may have hanging in a closet. Gather all your foraged and dried elements and lay them out on a table where you can view each piece. Don't be afraid to touch the items slowly and study their textures. Flip them around and twist them, considering each possible angle. Invite yourself to shut your mind off.

Next, make small bunches by gathering four or five dried flower stems or branches together and wrapping them tightly with floral twine. You could wrap each individual bundle on the metal ring as you go or make five or six bunches and play around with the location on the ring before securing them into place. I often use two floral twines during this process, one to wrap individual bunches and the other to secure the bunches onto the ring.

Keep in mind, things will slide on the ring! This style of wreath can be a wonderful lesson in letting go. You may place everything just so while the ring lies flat on the table, only to find that the moment you hang it on a wall everything shifts. Ask yourself, as the flowers shift, is the wreath any less beautiful? Of course not, it is simply different than you intended it to be. Recognize how tight you grip onto wanting things to be just so and as you intended. Are you able to let go a bit, and consider a different perspective?

For a final touch, I always add a bit of natural dyed ribbon or a special item to make the wreath unique. This might be dried seaweed, a seashell, or a small golden bell. There is no right or wrong when it comes to wreath-making. Just pick items that speak to you, and you'll be sure to create something you love. Don't forget this style of wreath will constantly evolve as you add bits of new organic elements all season long.

The Wild Mantle

The winter season is perfect for a bit of wilding. I usually dedicate an entire afternoon in mid-December to sourcing and creating a wild organic mantle above my fireplace. To begin, bundle up, then grab your favorite pair of snips and a tote bag. Take a deep breath and encourage yourself to focus only on the task at hand. Put aside your shopping and to-do lists for just a small bit of time. Search for elements in your backyard or someone else's, should they grant you permission. Perhaps look for long, interestingly shaped branches that can frame the mantle and act as the base of your design. Consider cutting grasses, aster, sumac, evergreen branches, or anything in the landscape that catches your eye. At this point in the season, most wild items that can be harvested are golden or evergreen.

When you return home, begin by securing branches together with twine, and anchor these items to the mantle with a small hook or nail. When you have created a solid base, build upon the lines by adding greenery and wrapping it loosely with twine. Next weave in bits of dried flowers, fruits, seaweed, shells, and/or berries. I like to add whole items like oranges and pomegranates. Consider the rule of three when bunching together smaller items. As you build up your design, step back from time to time to take in the overall shape. Often we are prone to overdoing things, but with a wild mantle less is certainly more. Make sure the lines of the original branches are still visible, otherwise the plot

can become lost. As this can be a big project, do leave the room and enter again, and again. This will allow you to see the design with new eyes. Just as with your wreath, a wild mantle doesn't have to be a final finished product. Continue to add to your design as the winter progresses.

The Indoor Garden

While it is easy to assume that your days with fresh flowers are over until spring, I would ask you to consider bringing flowering winter bulbs into your home. I've noticed that flowering bulbs have mostly gone out of fashion, replaced by an unfaltering love of houseplants, such as the fiddle-leaf fig. To me this is unfortunate, as winter bulbs add a lovely organic element to your home. And, just like most things I enjoy, they are incredibly easy to grow. A collection of flowering bulbs can provide you with fresh flowers until your snowdrops emerge in very early spring.

Pretty much every bulb can be forced in winter. Perhaps the most popular are paperwhites but personally I cannot stand their smell. However, there are many other flowers you can play around with when it comes to winter forcing. A collection of bulbs in different-size pots can be an eye-catching bit of home decor. They are also lovely to admire in the early-winter sun, when the world feels so cold and dark. Here are some things to consider when forcing bulbs indoors.

WATER

The most important bit of information for forcing bulbs is to be sure not to saturate the bulb in water as it will rot. The small white roots at the bottom of the bulb are what you want to keep watered.

PLANTING MEDIUM

There are three possible methods for growing bulbs:

» Place bulbs in a pot of small stones, gently resting the bulb on top of the stones.
» Place bulbs in well-drained soil, with roughly three-quarters of the bulb covered.
» Keep bulbs only in water and search out vases designed for bulb forcing; these often look like an hourglass.

TEMPERATURE

Determine if your bulbs need a chill period; many do, but some, such as amaryllis, do not. For bulbs that require a chill period, make sure they are kept in a room that is 45 to 50 degrees with minimal sunlight. Many people will start their bulbs in a garage, a refrigerator, or in a cooler. Once greens have popped out of your growing medium, you can bring them to temperature and into the light.

My Favorite Bulbs for Forcing

» **Amaryllis** (*Amaryllis* spp.)**:** As these bulbs and flowers are quite large, I usually pot them up with soil and add a layer of stones on top. This enormous bloom does not require a chilling period, meaning you can place them in a bright, warm, sunny location in your home from the start. Rotate the pot to prevent the flower from arching toward the sun in one direction. These are long bloomers, and you can extend the bloom period by moving the flower to a cooler location once it has bloomed, which occurs in roughly ten to twelve weeks.

» **Daffodils:** Daffodils require a chilling time of twelve to sixteen weeks. As they can grow quite tall, plant the bulbs under soil until the tip of the bulb is just barely sticking out; consider adding small stones to hold it in place. Gradually pull them into a room that is 50 to 60 degrees with low light for one to two weeks until shoots emerge. Finally, place them in a warmer room and direct sunlight.

» **Fritillaria:** These beautiful checkered flowers require a chill period of twelve to sixteen weeks. After the roots are well-established and green growth begins to show, bring them inside to bloom. Keep in sunlight, but cool temperatures indoors will hold the bloom period longer.

» **Hyacinth:** I enjoy forcing hyacinth in water in an old-fashioned Victorian forcing vase. Hyacinth requires a chilling period, so keep it in a location just above freezing and in the dark initially. The bloom will take roughly thirteen weeks to flower.

» **Muscari:** Also known as grape hyacinth, these delicate little purple blooms put out a lovely fragrance. Like hyacinth, they require a chilling period of roughly ten weeks. When the blooms begin to emerge, bring them out of the chilling room and put them into direct sunlight.

» **Tulips:** Tulips require a lengthy chilling period of roughly twelve to nineteen weeks; keep them in paper bags in a refrigerator, away from fruit. After the chill period, pot them up in soil, covering the tip of the bulb with roughly one inch of soil. Store in a dark, cool place for an additional one to two weeks. Once little bits of growth emerge, place in a room with indirect sunlight that is roughly 60 to 70 degrees. Water to keep the soil moist but be sure not to get the soil wet, as this can cause rot.

A Time for Reflection

As the year closes, but before one begins the cycle over again, take a moment to reflect upon your year in the garden. In short order you will be opening seed catalogs, counting the inches between rows of seedlings, and ordering dahlias. The dreaming state of the garden is sure to create big, swirling bands of color, even in the darkest of nights.

In the coming months it will be important to review your notes from this season. Be sure to dust off your succession planting chart and make modifications. Some of the flowers listed will prove to be lofty goals on paper, never realized. But other blooms will seem so real when you recall them to mind, you can almost feel and smell them between your fingers.

Begin to identify core flowers that were easy to grow and brought you joy, and then consider trying a few new varieties. Ask yourself again, what is the purpose of my garden? What do I want to bring to myself and my community by digging in the dirt? The answer may change from year to year. Remember, there is no shame in dialing back or doing less.

This winter, order in seed catalogs to browse, drink tea slowly while dreaming of waist-high cosmos, and hold near the memory of those who taught you to grow or love flowers. Borrow a library book full of images of faraway gardens. Lay low and dream big. And when the ground thaws, take a deep breath and begin again.

> Soon will set in the fitful weather, with fierce gales and sullen skies and frosty air, and it will be time to tuck up safely my Roses and Lilies and the rest for their long winter sleep beneath the snow, where I never forget them, but ever dream of their wakening in the happy summer yet to be.
>
> **—Celia Thaxter**

References

Books

Benzakein, Erin. 2017. *Floret Farm's Cut Flower Garden*. San Francisco: Chronicle Books.

Booth, Abigail. 2019. *The Wild Dyer: A Maker's Guide to Natural Dyes with Projects to Create and Stitch*. New York: Princeton Architectural Press.

Burke, Nicole Johnsey. 2020. *Kitchen Garden Revival: A Modern Guide to Creating a Stylish, Small-Scale, Low-Maintenance Edible Garden*. Beverly, MA: Quarto Publishing Group.

Elworthy, Bridget, and Henrietta Courtauld. 2023. *The Land Gardeners: Cut Flowers*. New York: Thames & Hudson.

Geall, Christin. 2020. *Cultivated: The Elements of Floral Style*. New York: Princeton Architectural Press.

Haller, Rebecca, Karen Kennedy, and Christine Capra. 2019. *The Profession and Practice of Horticultural Therapy*. Boca Raton: CRC Press.

Leopold, Aldo. 2020. *A Sand County Almanac*. New York: Oxford University Press.

Merrick, Amy. 2019. *On Flowers: Lessons from an Accidental Florist*. New York: Artisan.

Palmer, Nigel. 2020. *The Regenerative Grower's Guide to Garden Amendments: Using Locally Sourced Materials to Make Mineral and Biological Extracts and Ferments*. White River Junction, VT/London: Chelsea Green Publishing.

Partridge, Bex. 2020. *Everlastings: How to Grow, Harvest and Create with Dried Flowers*. London: Quadrille Publishing.

Perenyi, Eleanor. 2002. *Green Thoughts: A Writer in the Garden*. New York: The Modern Library.

Proust, Milli. 2022. *From Seed to Bloom: A Year of Growing and Designing with Seasonal Blooms*. London: Quadrille Publishing.

Redwood, Ark. 2011. *The Art of Mindful Gardening: Sowing the Seeds of Meditation*. East Sussex: Leaping Hare Press.

Siegfried, Rachel. 2023. *The Cut Flower Sourcebook: Exceptional Perennials and Woody Plants for Cutting*. London: Filbert Press.

Stuart-Smith, Sue. 2021. *The Well Gardened Mind: The Restorative Power of Nature*. New York: Scribner.

Thaxter, Celia. 2001. *An Island Garden*. Boston: Houghton Mifflin.

Wall Kimmerer, Robin. 2013. *Braiding Sweetgrass: Indigenous Wisdom, Scientific Knowledge, and the Teachings of Plants*. Minneapolis: Milkweed Editions.

White, Katharine S. 2015. *Onward and Upward in the Garden*. New York: NYRB Classics.

Young, Damon. 2012. *Philosophy in the Garden*. Melbourne University Press.

Websites

Annie White Research
pollinatorgardens.org/2013/02/08/my-research/

Emily Dickinson's Herbarium
library.harvard.edu/collections/emily-dickinson-collection

Gardener's Workshop
thegardenersworkshop.com

Growing Kindness Project
growingkindnessproject.org

Solarization and Occultation Practice
extension.umn.edu/planting-and-growing-guides/solarization-occultation

Sustainable Floristry Network
sustainablefloristry.org

The Wild Seed Project
wildseedproject.net

Acknowledgments

This book would not exist without the beautiful photography of Lindsay Fairchild, whose generous smile and willingness to work with ever-fluctuating floral bloom windows was integral to telling the story of a beginner's garden. Special thanks to Katie Rocheford for creative direction in the early stages of this project and Janet Blyberg for not only styling every last bit of dirt but lending us her beautiful vase collection. I am indebted to the Old York Historical Society for graciously sharing their interior space to create beautiful interior photographs. Thank you to Blythe Armitage, my sister, for editing my writing since you gained the ability to read.

To my husband and children, thank you for meeting me in the garden, every morning and every evening, keeping me company, and working alongside me. I am forever grateful to my dear friends who encouraged me to follow my path toward flowers. These women were always first in line to buy flowers, sign up for a workshop, or remind me I could write and had something worthwhile to say. A nod to the talented group of women flower farmers in Maine who have been incredibly generous with their knowledge over the years. If you don't yet know your local flower farmer, please take the time to make their acquaintance. To Makenna Goodman and Timber Press for recognizing the importance of encouraging beginner gardeners to grow flowers.

And last, but not least, thank you to my family who lived at the little white house with the red door on Raynes Neck Road. A place full of poetry, music, dirt, and nature. A childhood with space to roam and question the natural world, alongside a brilliantly clever brother and sister.

Index

Achillea millefolium, 153
Achillea spp., 41
ageratum, 120
Ageratum spp., 43
agribon, 184, 185
Alcea rosea, 38
Alchemilla mollis, 39
algae, 200
amaryllis, 218
Amaryllis spp., 218
American Daffodil Society, 98
Anderson, Hans Christian, 97
Anemone spp., 183
Anethum graveolens, 51
'Angélique' tulip, 102, 103
annuals, 32, 42–49, 55, 175, 180
    harvesting, 120–121
    succession planting, 57–59
Antirrhinum spp., 48
aphids, 121, 126, 143–144, 145
Appledore Island, 12
apple mint, 52
Appleton, Jay, 66
apps, identification, 75, 78
'Apricot Whirl' daffodil, 99
Aristotle, 157
arrangements. *See* floral design
arrowwood, 106
Asclepias tuberosa 'Hello Yellow', 152
aster, 175, 198, 215
astrantia, 118, 198, 202
Astrantia major, 34
Atkins, Anna, 200

backpack sprayers, 86–87
balsam fir, 211
bamboo stakes, 81
basil, 51
beans, 67
bearded iris, 32, 118

bees, 150–151, 152
beginner's mind, 12, 74
Benzakein, Erin, 26, 181
berries, 210, 211
biennials, 32
biodiversity, 150
birds, 144, 208
bittersweet, 211
blackberry, 53
black-eyed Susan, 34, 118, 150, 153
'Black Parrot' tulip, 102, 103
blanketflower, 35, 120
blueberry, 53
Bluetooth timers, 87
bouquet dill, 51
bouquets, 28–29
bowls, 134, 135
boxwood, 20, 31, 211
British Algae (Atkins), 200
broad forks, 80
buckets, 80, 84, 114, 130, 190
bulbs, 96–105, 216–218
buttercup, 20
butterflies, 151, 152

'Cafe au Lait' dahlia, 175
calendula, 54, 68, 120, 150, 157, 159
Calendula officinalis, 43, 56
Callistephus chinensis, 181
cedar, 211
Celosia spp., 44
Chai, Julie, 26
Chamaemelum nobile, 159
chamomile, 159
Chasmanthium latifolium, 153
checkered lily, 97
Cherokee, 67
chickadee, 180
chicken wire, 130–131, 136–137
China aster, 181

Chinese forget-me-not, 56
clay, 85
cleaning tools, 190–191, 192
closing your garden, 179–180
cockscomb, 44, 120, 202
Coleman, Eliot, 61, 84
color palette, 30, 32, 100
community connection, 155–157
companion planting, 67–68, 145
compost, 91, 95–96
coneflower, 35, 118, 150, 153
Consolida spp., 56
container gardens, 66, 85, 89–90
Coreopsis 'Red Satin', 41
Coreopsis spp., 41
Coreopsis tinctoria, 41
Coreopsis verticillata 'Moonbeam', 41
corn, 67
cosmos, 29, 55, 90, 120, 198
Cosmos spp., 44
Cotinus spp., 107
Courtauld, Henrietta, 26, 51, 104
creeping thyme, 89–90
crop rotation, 104, 145
crown gall, 187
cultivars, 152
Cultivated (Geall), 26
cultivators, 80
cut-and-come-again flowers, 70, 90, 112, 142
cutting back, 180
cyanotypes, 200
Cynoglossum amabile, 56

daffodil, 74, 97, 98–100, 115, 118, 140, 218
dahlia, 31, 100, 104, 111, 126, 136, 155, 220
    companion planting, 67
    digging, 185–187
    dividing and storing, 187–190
    harvesting, 120
    late-season blooms, 181, 183

Dahlia spp., 45
Dakota, 149
daylily, 180
deer, 100, 111, 144
delphinium, 31, 118
Delphinium spp., 36
design aesthetic, 28–29
designing the garden, 66–67
Dickenson, Emily, 197
Dictyota dichotoma, 200
Diervilla lonicera, 107
Digitalis spp., 36
dirty flowers, 100, 115
donations, 155
Dowding, Charles, 90
'Dream Touch' tulip, 102, 103
dried flowers, 195, 201–208
drip irrigation systems, 86–87

eastern white pine, 211
echinacea, 32
Echinacea purpurea, 153
Echinacea spp., 35
Echinops spp., 37
edible flowers, 54, 158–159
electric tools, 84
elevated beds, 90
'El Niño' tulip, 102, 103
Elworthy, Bridget, 26, 51, 104
Emilia sonchifolia var. *javanica* 'Irish Poet', 49
ergonomic tools, 84
Eustoma spp., 46
Eutrochium spp., 153
evergreen, 210, 215
everlasting flowers, 201–202
Everlastings (Partridge), 26

fabric dying, 161–165
fern, 198, 200
fertilizer, 70, 146

INDEX | 227

feverfew, 28, 31, 52, 119
filler flowers, 31
fish fertilizer, 146
floral brick, 134, 135
floral clay, 81
floral design
    in darkness, 140–142
    design aesthetic, 28–29
    dried flowers, 203–208
    setting the stage, 134, 136–137
    sustainable, 129, 211
    tips, 138–139
    tools, 130–131
    vessels, 133–134
    winter, 211
    workspace, 128, 130
floral frogs, 130, 136–137
floral knives, 130
floral recipes, 70
floral shears, 114
Floret Farm's Cut Flower Garden (Benzakein and Chai), 26, 181
flowers
    annuals, 32, 42–49
    cultivars, 152
    cut-and-come-again, 70, 90, 112, 142
    dirty, 100, 115
    dried, 195, 201–208
    edible, 54, 158–159
    everlasting, 201–202
    identification apps, 75, 78
    late-season blooms, 181, 183
    memories of, 20–21
    native, 150–153, 175
    perennials, 32–41
    pressed, 197–199
    sourcing, 54–55
    types of, 30–31
On Flowers (Merrick), 26
focal flowers, 28, 31, 138–139
foliage, 31
food pantries, 155
forcing bulbs, 216–218
forsythia, 74
Fort Snelling, 149

foxglove, 29, 31, 32, 36, 119
French marigold, 54
fritillaria, 218
Fritillaria, 97
Fritillaria meleagris, 97
Fritillaria persica, 97
frost cover, 185
frost dates, 23, 108, 175, 183
'Frosted Explosion' grass, 121, 202
fruit, 53

Gaillardia spp., 35
Galanthus, 97
galvanized electrical conduit, 185
garden, word origins, 157
garden books, 26
garden gloves, 80
garden hoes, 80
garden hoops, 184–185
gardening carts, 81
garden of the mind's eye, 20, 21, 68
garden totes, 114
garlands, 208
Geall, Christian, 26
germination, 59
giving away flowers, 155
globalization, 201
globe amaranth, 45, 121, 202–203
globe thistle, 37
goldenrod, 37, 119, 150
Gomphrena spp., 45
grape hyacinth, 97
grasses, 33, 215
greenery, 31, 139, 210
greenhouses, 54–55, 95
Green Thoughts (Perenyi), 26
groundhogs, 100, 121, 143, 144
grounding practices, 95, 122–123, 210
Growing Kindness Project, 155
growing space, observing, 27
growing zones, 23

haircombs, 205
hardening off process, 108–110

harvesting, 69, 70, 112, 180
   annuals, 120–121
   bulbs, 118
   how to, 115–116
   perennials, 118–119
   timing considerations, 117
   tools, 114
   woody perennials and shrubs, 120
Hassam, Childe, 100
healing gardens, 14
herbarium-inspired pressed flowers, 197
herbs, 31, 50–53, 210
hip bags, 114
hoes, 80
holly, 210
hollyhock, 22, 32, 38, 119, 144
holsters, 114
hoops, 184–185
hori hori knives, 80
Houghton Library, Harvard University, 197
hyacinth, 74, 97–98, 118, 218
hyacinth bean, 46, 150
Hyacinthus spp., 97
hybrids, 149
hydrangea, 67, 180, 198, 211
Hydrangea arborescens 'Annabelle', 180
Hydrangea paniculata 'Limelight', 180
Hydrangea spp., 106

'Ice Follies' daffodil, 99
indoor gardens, 216–218
inspiration, 21–23, 54, 70
Integrated Pest Management (IPM), 144
interplanting, 67–68
invasive plants, 211
iris, 180
'Irish Poet' tassel flower, 121
Iris spp., 38
Iroquois, 67
irrigation, 86–87
An Island Garden (Thaxter), 12, 14, 26
ivy, 210

Japanese anemone, 183
Japanese beetles, 145

Japanese hand hoes, 80
Joe-pye weed, 153
jute twine, 81, 131

killing frost, 183–184
Kitchen, Deanna, 155
kneelers, 84
Korean natural farming, 148
kraft paper wrap, 81

Lablab purpureus, 46
Lady Bird Johnson Wildflower Center, 152
lady's-mantle, 39, 119
The Land Gardeners (Elworthy and Courtauld), 26, 104
larkspur, 56
late-season blooms, 181, 183
Lathyrus odoratus, 57
Lavandula spp., 39
Lavandula x *intermedia* Phenomenal, 39
lavender, 39, 54, 119, 159, 180
lazy Susan, 130
leaf stripping, 116
'Lemonade' cosmos, 32
lemon balm, 159
lilac, 20, 54, 106, 111
lily, 33
lisianthus, 46, 68, 121
loam, 95–96
lobster shell extract, 148
loneliness, 157
long-handled wire weeders, 84
loppers, 81
love in a mist, 56
lowbush blueberry, 106
Lower Sioux Reservation, 149
lupine, 143–144
Lyceum, 157

mantles, 215–216
marigold, 100, 163, 183
mass flowers, 31
meditation, 167–168

Melissa officinalis, 159
memory, 20–21, 140
mental health benefits, 14–15, 210
Mentha suaveolens, 52
Merrick, Amy, 26
micronutrients, 93
mildew, 180
milkweed, 211
mind's eye, 20, 21, 68
mint, 52
mistletoe, 210
molasses, 146
Monarda fistulosa 'Claire Grace', 152
Monarda spp., 153
moon walks, 168
morning meditation, 167–168
mountain mint, 52
mulch, 180
muscari, 218
Muscari spp., 97

Narcissus spp., 98
nasturtium, 54, 90, 150, 159, 198
Nasturtium spp., 159
nativars, 152
native flowers, 150–153, 175
natural dyes, 161–165
Neolithic revolution, 157
New England aster, 150, 151
New York Botanical Garden, 33
nigella, 29, 121, 150, 203
Nigella damascena, 56
Nigella spp., 47
ninebark, 106
nitrogen, 67
no-dig method, 90–91
nonjudgment, 138
northern bush honeysuckle, 107
northern sea oats, 153
no-till tulip-bed method, 105

observation, 27, 170–171, 179
occultation, 192, 193, 195
Ocimum spp., 51

Onward and Upward in the Garden (White), 20, 26
opening the garden gate, 157
'Orca' tulip, 102, 103
Organic Materials Review Institute, 144
organic matter in soil, 93
ornamental gardens, 157
ornamental grasses, 31
oshibana pressed flowers, 197
Oudolf, Piet, 33

Paeonia spp., 40
Palmer, Nigel, 148
'Palmyra' tulip, 102, 103
Panicum capillare 'Frosted Explosion', 47
Papaver rhoeas, 56
paperwhite, 216
'Parrot King' tulip, 102, 103
Parthenium spp., 52
Partridge, Bex, 26
peony, 33, 40, 119, 121, 136, 168, 180
perennials, 32–41, 111, 175, 180, 181, 183
 harvesting, 118–119, 120
Perenyi, Eleanor, 26
pests, 143–145
pH levels, 93
phlox, 22
phlox paniculata, 153
Physocarpus spp., 106
pinching, 112
pincushion flower, 40, 119, 203
pine, 210
pitchers, 134, 135
plant supports, 81
plastic flowers, 201
pollinators, 33, 67, 150–151, 152
Polygonatum spp., 41
Poppies, Isles of Shoals (Hassam), 100
poppy, 128, 200
posy, 203–204, 206
pots, 89–90
preparing your garden, 89–96
present-moment awareness, 121–123
pressed flowers, 197–199

'Professor Einstein' daffodil, 99
prospect-refuge theory, 66
Proust, Milli, 26
pruning, 180
Pycnanthemum spp., 52

Queen Anne's lace, 31, 198, 200
'Queensday' tulip, 102, 103

radish, 68
rain gauges, 86
raised beds, 85, 90, 95
rakes, 81
raspberry, 50, 53, 142, 175
ratchet pruning shears, 84
red clover, 159
reflection, 220
The Regenerative Grower's Guide to Garden Amendments (Palmer), 148
Rhus typhina, 107
rhythm, creating, 68–70
ribbon, 81
rose, 31, 33, 54
rudbeckia, 32
Rudbeckia hirta, 153
Rudbeckia spp., 34

saving seeds, 147, 149–150
Scabiosa spp., 40
schedules, 57–59, 68–70
sculptures, 205
season extension, 183–185
seaweed, 147, 211
seaweed fertilizer, 146
secondary flowers, 31
seedlings, hardening off, 108–110
seeds
　native blooms from, 152
　saving, 147, 149–150
　soil blocking, 61–63
　sowing, 56–57, 69
　starting, 55, 60–61
　succession planting schedules, 57–59
From Seed to Bloom (Proust), 26

shears, 81, 84
Shirley poppy, 56
shrubs, 105–107, 120
silver tip, 57
'Sir Winston Churchill' daffodil, 99
size, garden, 88
slow gardening, 167–171
slugs, 143, 144
smoke tree, 107
snapdragon, 29, 31, 32, 48, 57, 68, 100, 121
snips, 114, 130
snowdrop, 96, 97
soil
　amendments, 93, 145–147, 148, 179
　blocking, 61–63
　controlling, 92
　dryness testing, 84–85
　good, 92
　mason jar test, 85
　mixes, 95
　obtaining, 95–96
　readiness, 56
　testing, 92–93
solarization, 192
Solidago spp., 37
Solomon's seal, 41, 119, 168, 175
sourcing flowers, 54–55
sowing seeds, 56–57, 69
spades, 81
spigot splitters, 87
sqh, 67
staghorn sumac, 107
The Starry Night (van Gogh), 36
starting seeds, 55, 60–61
starting small, 88
stock, 68
strawflower, 48, 121, 203, 204, 208
stretching, 168, 170
succession planting, 57–59
sumac, 211, 215
sunflower, 31, 67, 69, 126, 150, 157
sunlight, 27, 66
sunprints, 200
supplies, 78–84, 114, 130–131, 184–185, 188, 190–191, 192

sustainability, 129, 211
Sustainable Floristry Network (SFN), 129
Suzuki, Shunryū, 74
sweet pea, 57
Syringa spp., 106

Tagetes spp., 183
taking stock, 181
tarping, 192
tassel flower, 49
Thaxter, Celia, 12, 14, 15–16, 20, 21, 22, 26, 55, 74, 78, 100, 115, 126, 133, 220
therapeutic horticulture, 14–15
three sisters method, 67
Thumbelina (Anderson), 97
tickseed, 41, 119
tools, 78–84, 114, 130–131, 184–185, 188, 190–191, 192
toxicity, 100
traditions, 208–211
transplant shock, 108
trench method, 105
Trestrail, Charlene, 129
Trifolium pratense, 159
tripod sprinklers, 86
Triticosecale spp., 57
trowels, 81
tubers, 111, 186–190
tulip, 31, 97, 100–105, 118, 218
tussie-mussie, 203–204
twine, 81, 131

Vaccinium angustifolium, 106
van Gogh, Vincent, 36, 38
vases, 133, 134, 135
vegetables, 155
Veronicastrum virginicum 'Lavendelturm', 152
vessels, 133–134, 135
Viburnum spp., 106
victory gardens, 11
violet, 198

walkabouts, 157
watering cans, 86
watering plans, 84–87
water sources, 27
water wands, 86
weeding, 69, 70, 84, 126, 179–180, 195
weekly rhythm, 68–70
wheelbarrows, 81
White, Annie, 152
White, Katharine S., 20, 26
white spruce, 211
wild bergamot, 153
wild geranium, 150
wildlife, 100, 111, 143, 144, 180, 211
wild raspberry, 31
Wild Seed Project, 152
willow tree, 161
worm farms, 147
wreaths, 210, 212–213

Xerochrysum bracteatum, 48

yarrow, 28, 31, 32, 41, 100, 119, 153, 159, 168, 203

Zen Mind, Beginner's Mind (Suzuki), 74
zero-waste floristry, 129
zinnia, 55, 67, 69, 90, 100, 115, 121, 143, 150, 157, 208
Zinnia elegans 'Zinderella Peach', 49
Zinnia haageana Jazzy Mixed, 49
Zinnia Oklahoma Series, 49
Zinnia spp., 49

Elizabeth Brown is the founder of Foxglove Farmhouse, a cut flower garden from which she grows flowers for her community and educates others to do the same. She lives in Maine with her husband and two children. This is her first book.